GAS: NATURAL ENERGY

GAS:
NATURAL ENERGY

Richard Cassidy

FREDERICK MULLER LIMITED
LONDON

First published in Great Britain 1979 by
Frederick Muller Limited, London NW2 6LE

British Library Cataloguing in Publication Data

Cassidy, Richard
Gas.
I. Gas, Natural
I. Title
333.8'2   HD9581.A2

ISBN 0-584-62056-X

Set IBM 11pt Baskerville by 𝐀\ Tek-Art, Croydon, Surrey.
Printed in Great Britain by
The Garden City Press Limited
Letchworth, Hertfordshire SG6 1Js

CONTENTS

PHOTOGRAPHS

Photographs courtesy of the Public Relations Department, British Gas Corporation.

PREFACE

The need for an up to date book about natural gas and the gas industry in Britain became apparent to me in the course of my job as information officer for the British Gas Corporation. I have often been asked by school pupils and college students for information which they could use in projects on gas, the North Sea, or energy in general. Teachers also ask if there is some comprehensive factual source which will give them a straightforward introduction to a complicated subject, touching on such diverse academic disciplines as chemistry and history, economics and geology. Similarly, people applying for jobs or starting work in the gas industry sometimes want to know how their specialized occupation relates to the total picture, from exploration in the North Sea to repairing gas fires. For several years, no one book has been available to meet these needs.

I have tried to write an objective book about gas and the gas industry, for readers with no specialized knowledge of the subject, in order to fill this gap in the available information. I wrote it in my own time, and confined myself to facts rather than opinions as far as possible. Nevertheless, I must make it plain that although I work for British Gas, the responsibility for any opinions expressed and the work as a whole is mine alone.

I would like to thank my colleagues and friends at British Gas, who have taught me all I know about gas, and Michele, who made so much coffee and exercised so much patience while I was writing this book.

INTRODUCTION:
FROM THE NORTH SEA
TO OUR HOMES

Far out in the North Sea, hundreds of men are working on steel and concrete platforms, lashed by gale-force winds and soaked by the spray from stormy waves. Meanwhile, millions of families sit in warm homes, eating their dinners, without ever having to concern themselves with the way in which the heat they need is brought to them. We all depend on plentiful, reliable supplies of fuel for comfort and hot food at home. We also depend on the fuel used in almost every industrial process for the prosperity of the country as a whole. Energy is central to the well-being of the economy, but it also has an immediate relevance for all of us, the importance of which is easy to appreciate in our daily lives. It is impossible to imagine how we could manage without the fuel which we use all the time without a second thought — gas.

The arduous, if well-paid, lives of workers on North Sea platforms may seem remote from the daily existence of the majority of the population. In fact, they are so closely linked that each group depends on the other. Our way of life demands the consumption of large quantities of fuel, which they are producing. Without us, the consumers, there would be no incentive to undertake the development of the gas and oil found beneath the North Sea. The physical link between the platforms and Britain's homes and factories is provided by the gas industry. This book describes how gas reaches our homes from the North Sea, what it is, where it is found, and how it is used, as well as taking a look at what might happen to gas in the future.

We can readily understand the importance of gas in our own lives, if, like most people in Britain, we use gas at home for heating and cooking. Perhaps less widely known is the importance of gas to Britain as a whole. Some 14 million homes have a gas supply; about 11 million use gas for cooking; 5 million have gas central
10

heating. In all, 45 per cent of all the energy supplied to British homes is provided by gas, making it far and away the most popular domestic fuel. This proportion is growing, as more and more families install central heating. More than half the homes in Britain now have central heating, and more than 70 per cent of the new central heating systems being bought use gas.

The statistics confirm the impression gained by simple unscientific observation that gas is the most popular domestic fuel, because it is clean, economical and reliable. Outside the home, in industry and in commercial applications like offices and hotels, gas also plays a major role as a source of heat. In all, gas supplies 20 per cent of the primary energy demand in Britain.

Nearly all of this gas comes from the North Sea. Not so long ago, all the gas used was manufactured from coal or oil. It was not until 1967 that the first natural gas from the North Sea was brought ashore, but ten years later, in 1977, the programme of converting the whole gas supply system and all gas-burning appliances to use natural gas was completed. The discovery and development of the natural gas reserves in the North Sea not only led to the closure of the gas works, but also permitted a rapid expansion in the amount of gas available. Common sense as well as statistics show how important natural gas now is as a major source of energy for Britain. Without it, we would still be dependent on importing oil from abroad to provide this energy in order to maintain the same standard of living. The Government estimated that natural gas from the North Sea thus saved Britain £2,000 million a year on the balance of payments.

The importance of gas as a fuel makes the gas industry one of the crucial public services on which the British economy depends, but the rapidity of its transition and growth since the discovery of natural gas have made it less well known than it deserves. Most people probably know that they are now supplied with natural gas from the North Sea, but may not be aware of the size of gas's share of the energy market, or of the fact that British Gas is among the ten largest businesses in the country.

Similarly, the process whereby natural gas is produced and brought ashore and supplied to customers is little known. Like the other vital public services, gas supply is taken for granted, at least until something goes wrong. We never give a thought to the fact that we can get as much water as we want just by turning a tap, rather than having to carry it a bucket at a time from a well, unless a drought leads to restrictions on water usage. The good record of the gas industry in maintaining continuous supplies to its cus-

11

tomers has meant that they have only occasionally been interrupted, either by failures in the system or by industrial action, and then such problems have only been brief and local. There is seldom any stimulus for the customer to concern himself with his gas supply, apart from paying for it. Gas is always there, to provide as much heat as we want, whenever we want it.

The system of gas supply which ensures that gas can be relied upon to provide a continuous source of heat is the subject of this book. It is the link between the unfamiliar and exotic world of North Sea exploration and the mundane everyday matters of warm homes and hot water. To begin at the very beginning, we have to find out what natural gas actually is, and how it was formed hundreds of millions of years ago.

1

NATURAL GAS – ORIGINS AND CHARACTERISTICS

How natural gas was formed
Over 300 million years ago, shallow seas and swamps covered much of the area we now know as north-west Europe. The climate was hot and humid, so primitive plants flourished in the marshes, particularly in a low-lying basin which stretched across what is now the southern part of the North Sea. In this marshy basin, generation after generation of plants grew, died and decayed, building up a layer of dead vegetable matter. At the same time, minerals were being washed away from nearby areas of higher land, carried down to the basin and deposited as a sediment. The dead plants and sediment were gradually buried and compressed, to form coal deposits.

Geologists know this era of swamps, when coal was formed, as the Carboniferous age. As the climate changed, to become hot and dry, it gave way to the Permian age. The swamps dried up, and in this desert climate the wind eroded grains of sand from higher land to the south and deposited the sand in the basin. The layer of coal was thus covered by a layer of sand, which became sandstone, about 270 million years ago. Later in the Permian age, about 225 to 250 million years ago, a shallow sea, known as the Zechstein Sea, covered the area. In this arid climate, lakes of salt water were formed, cut off from the sea, as the water evaporated. Further evaporation dried up the lakes, leaving behind thick layers of salt.

The basin which is now the southern North Sea thus came to have a layer of coal beneath a layer of sandstone beneath a layer of salt. For further hundreds of millions of years, this basin, and others like it, continued to accumulate sedimentary deposits. When they were covered by the sea, minute sea creatures like plankton, bacteria and simple plants, died and settled to the sea bed. Here they were mixed with mineral particles eroded from

13

areas of land, carried down to the sea by rivers, and buried by further layers of sediment. The pressure acting on the buried deposits of sediment compacted them into rocks such as marine shales. Occasionally, when the right conditions of temperature and pressure were present, bacterial action on the buried sea creatures in association with mineral catalysts in the inorganic sediment produced hydrocarbon molecules. Similarly, chemical reactions in the layers of coal under the weight of further deposits sometimes resulted in the formation of hydrocarbons.

Hydrocarbons are, as their name would imply, compounds of hydrogen and carbon in varying proportions. There are many different hydrocarbons, some gaseous and some liquid, formed by chemical reactions in dead and buried organic matter such as coal and marine shales. One thing they have in common is that they can be used as fuels, or refined to produce fuels. The liquid hydrocarbons can be referred to broadly as oil, the gaseous hydrocarbons as natural gas. Both oil and gas are mixtures of various hydrocarbons, but the most important constituent of natural gas is the simplest hydrocarbon compound, methane (CH_4).

Oil and natural gas formed in source rocks like coal are relatively light substances, and where possible will tend to rise into the layer of rock above the source rock. In the southern North Sea basin, as we have seen, the coal deposits lie beneath a layer of sandstone. Sandstone is a porous rock, which normally contains water in the pore spaces which account for up to 30 per cent of its total volume. Oil and gas are lighter than water, and will thus rise into the pore spaces of the sandstone, displacing water, and travelling upwards and sideways until they reach the surface and escape. Occasionally, their route is blocked by an impermeable rock, through which they cannot travel, and oil and gas are trapped. If the layer of impermeable rock is horizontal, the gas and oil will tend to migrate sideways beneath it until some escape route has been found, but when there is a geological formation restricting travel both vertically and laterally, the gas and oil are trapped. The porous rock containing hydrocarbons is known as a reservoir, the impermeable rock preventing their escape as a cap.

The geological history of the southern North Sea has been such that all the necessary requirements for the accumulation of gas in reservoirs are present. The source rocks where the gas was formed are the Carboniferous coal measures. The gas has migrated into the porous sandstone of the Permian age, and has been trapped beneath the impermeable layer of salt left behind by the evaporation of the salt water lakes more than 200 million years ago. This is the

14

situation in the southern North Sea basin. In the northern North Sea, oil and gas were formed somewhat later in the deposits of marine organisms, and trapped in the sandstone and chalk of the Lower Tertiary age, up to 65 million years ago.

Geological traps to contain oil and gas in reservoirs can be formed in several ways. The simplest is an anticline — where the layers of rock have folded to form a structure like a dome, gas will be trapped at the highest point of the dome, beneath the impermeable cap. An anticline can also be formed by the upward intrusion of salt into the rock layers above it. In such a salt dome, gas accumulates in the highest part of the porous rock which has been forced upwards by the thrust of the salt. A third type of trap may be formed by faulting — when there are sloping layers of porous and impermeable rocks, one above the other, a geological fault may cause the impermeable rock to move downwards relative to the porous. The impermeable rock then closes off the upward slope of the porous layer, and gas can be trapped. Another trap may be formed by sedimentation, depositing an impermeable layer on top of older porous rocks which have been tilted and eroded.

Whichever way the trap is formed, the principle is much the same so far as the accumulation of a reservoir of gas is concerned. The gas migrates from its source into porous rock, displacing the water which once occupied the pores, until an impermeable cap rock prevents it travelling any further. The gas collects in this trap, the cap above it and water beneath it in the porous rock. There it stays until a route for its escape through the impermeable layer is created by drilling a hole from the surface down to the reservoir.

The composition of natural gas

Both oil and natural gas are mixtures of hydrocarbons, and are often found together. Some gas fields, like those in the southern North Sea, contain little or no oil, and there are also some oil fields where there is little gas associated with the oil, but whether the gas is found by itself or in association with oil, its principal constituent is always methane. Every field is different, reflecting the processes which formed hydrocarbons from organic material in varying proportions, and when natural gas is produced from the field it brings with it water and light liquid hydrocarbons, known as condensate. The water, from the pores in the reservoir rock, and the condensate, have to be removed before the natural gas can be used, leaving a mixture of gaseous hydrocarbons and a few other gases, for supplying to customers who want to use natural gas as a fuel.

The principal constituent of natural gas is always methane (CH_4), and the most common of the numerous other hydrocarbons in the mixture are ethane (C_2H_6), propane (C_3H_8) and butane (C_4H_{10}). All these hydrocarbons are combustible, so they can be used as fuel. Natural gas can also contain incombustible gases which were produced together with the hydrocarbons, and in some cases sulphur compounds, which have to be removed in order to prevent air pollution when the gas is burned.

A typical example of natural gas, as used in Britain, would be the mixture produced from three fields in the southern North Sea and brought ashore at Bacton, on the Norfolk coast. When it has been treated to remove water and condensate, the composition by volume of this Bacton natural gas as used by gas consumers in Britain is:

	per cent
nitrogen	1.78
helium	0.05
carbon dioxide	0.13
methane	93.63
ethane	3.25
propane	0.69
butane	0.27
other hydrocarbons	0.20

It can be seen that Bacton natural gas contains a very small proportion of incombustible gases like nitrogen, helium and carbon dioxide, no sulphur compounds and no toxic gases like carbon monoxide. Methane is far and away the most significant component of the mixture. It is lighter than air — if air has a density of 1, the specific gravity of natural gas is 0.593.

The proportions of the various gases in natural gas can vary quite considerably from one field to another. Gas from the Groningen field in the Netherlands contains as much as 14 per cent nitrogen and only 81 per cent methane. Gas from Lacq in France has to be treated to remove large quantities of sulphur, and then contains only 0.30 per cent nitrogen and 97 per cent methane. Whatever the precise mixture, the most important fact about natural gas is that it can be used as a fuel.

Burning natural gas

Whether natural gas is used to cook a meal, heat a house or fire a furnace in a factory, the vital factor is that heat is produced when natural gas burns. The major constituent of natural gas is methane; when it burns, it reacts with the oxygen in the air to produce carbon dioxide and water vapour:

$$CH_4 + 2O_2 \rightarrow CO_2 + 2H_2O$$

Like other fuels, natural gas will only burn in the presence of sufficient oxygen — if there is too little or too much oxygen, combustion cannot take place. The limits of flammability, within which our typical natural gas will burn, are mixtures of between 5 and 15 per cent natural gas in air. The ideal mixture, in which there would be just the right amount of oxygen available in the air, would contain 9.25 per cent natural gas. This is known as a stoichiometric mixture — in this case, for each cubic metre of natural gas there would be 9.82 m³ of air.

As natural gas is only useful as a fuel because of the heat produced when it burns, it is important to know just how much heat can be produced, theoretically, from a given volume of gas. This is its calorific value, and it varies as the mixture of gases in natural gas varies. There is a different calorific value for the gas from each field, which makes it all the more important to know the calorific value; after all, the gas consumer is more interested in the amount of heat it will produce than its volume. Gas is therefore sold to customers on the basis of its heat content, rather than its volume, and the price is fixed at a rate of so much money for each unit of heat. In Britain this unit is the therm, but in SI the unit is the joule and its multiples. (The appendix contains more information about the units used and conversion factors.) The calorific value of our typical Bacton natural gas is 38.86 MJ for each cubic metre; Groningen gas, with its much lower methane content and more incombustible nitrogen, has a calorific value of 33.24 MJ/m³.

When natural gas burns in air, the reaction with oxygen produces heat, carbon dioxide and water vapour. The nitrogen and inert gases in the air and in the natural gas mixture are unaffected, so that stoichiometric combustion of one cubic metre of Bacton natural gas with 9.82 m³ of air would, in theory, produce 38.86 MJ of heat and 10.85 m³ of waste gases. These waste gases would have a composition of 9.67 per cent carbon dioxide, 18.66 per cent water vapour, 70.83 per cent nitrogen and 0.84 per cent inert gases.

Burning natural gas is rather like breathing, in that the oxygen in the air is used up, and carbon dioxide and water vapour are produced. We need to live in properly ventilated homes so that there is always a fresh supply of oxygen for us to breathe, and we are not suffocated by the replacement of oxygen by carbon dioxide and water vapour. In the same way, gas burners need a continuous supply of oxygen and some way to dispose of the products of combustion. The normal ventilation which ensures that we have oxygen to breathe and expels stale air is enough to ventilate appliances like cookers which burn small quantities of gas. Gas fires and central heating boilers use more gas, and thus need more oxygen, producing larger volumes of waste gases. Flues are needed to take away the products of combustion, and prevent the room filling up with carbon dioxide and water. The products of combustion are quite harmless in the sense that they are not poisonous, but it would obviously be as bad for our health to live in rooms where oxygen was displaced by these products, as it would be if there was no ventilation bringing us fresh air to breathe.

A more serious problem can result from a room filling with products of combustion when a gas appliance is working. There will be too little oxygen available to the burner, and incomplete combustion will result. Instead of non-toxic carbon dioxide and water vapour, the products of incomplete combustion due to a lack of oxygen can include carbon monoxide, which is very poisonous. Carbon monoxide is well known as being present in motor car exhaust gases, which have sometimes poisoned people foolish enough to run a car engine in a closed garage. In fact, when any fuel containing carbon, be it gas, petrol, coal, oil, coke or wood, is burned in the absence of sufficient oxygen, carbon monoxide may be produced. With gas appliances, incomplete combustion usually results from a lack of care in installing and maintaining them. Gas fires without flues, or with blocked flues, and water heaters without adequate flues to take away the products of combustion are responsible for a succession of cases of carbon monoxide poisoning, which could be avoided by regular checks on the safety and efficiency of gas appliances.

Liquefied natural gas

Liquefied natural gas, generally known as LNG, is exactly what its name implies — natural gas at such a low temperature that it changes from a gaseous to a liquid state. Just as steam becomes water, when its temperature falls below $100°C$, so natural gas becomes a liquid at temperatures below $-161°C$.

Like any other gas, natural gas occupies a much greater volume as a gas than as a liquid. LNG takes up only 1/600 of the volume occupied by the same quantity of natural gas under standard con-

18

ditions of temperature and pressure (15°C and 1013.25 mbar). It is this enormous reduction in volume that makes LNG so useful as a means of transporting and storing natural gas. There are obvious advantages in using just one tank to hold LNG instead of 600 conventional gas holders of the same size, which would be needed to store the same amount of gas in its gaseous state.

Natural gas is a mixture of gases which become liquids at different temperatures, and one of which, carbon dioxide, becomes a solid at a relatively high temperature — it is well known as dry ice. Before natural gas can be liquefied, it is necessary to remove carbon dioxide which would otherwise solidify and block pipes. The fact that natural gas is a mixture also means that LNG from different sources will have different characteristics, although methane is still the major component. An example of the characteristics of LNG might be provided by the LNG which is imported from Algeria. When it arrives in Britain, Algerian LNG has the following composition by volume:

	per cent
nitrogen	0.36
methane	87.20
ethane	8.61
propane	2.74
other hydrocarbons	1.09

This LNG has a boiling point of -161.3°C, under standard pressure. At this temperature its density is 468.7 kg/m^3 — in other words, its specific volume is such that one tonne of LNG occupies 2.134 m^3.

When the LNG is heated, one tonne of LNG produces 1,273 m^3 of natural gas at standard temperature and pressure — so the volume of gas is 597 times greater than the volume of LNG. The gas produced from this Algerian LNG has a density of 0.639 relative to air, and a calorific value of 42.62 MJ/m^3.

When Algerian LNG arrives in Britain, it is unloaded from the ship straight into storage tanks at the Canvey Island terminal in the Thames estuary. Elsewhere in Britain, natural gas is taken from the pipelines which carry it in its gaseous state and it is then liquefied for storage in tanks at sites near Glasgow, Manchester and Bristol, in south Wales and on the south bank of the Thames estuary. Generally, these LNG storage tanks have a capacity of 20,000 tonnes, so that when demand for gas increases in cold weather, some of the LNG can be regasified and fed back into the

19

pipelines, supplementing supplies from the North Sea. To give some idea of the value of storing natural gas as a liquid, a rough calculation shows that one 20,000 tonne LNG tank contains the equivalent of 25 million cubic metres of natural gas under standard conditions. In terms of heat, this is about 1,075 TJ. To put it another way, five such tanks contain LNG equivalent to all the gas consumed in Britain on an average day.

2

EXPLORATION AND DRILLING

Exploring for gas
As we have seen, sedimentary basins are depressions in the earth's crust where layer upon layer of deposits have accumulated. As the climate changed and continents and seas formed and disappeared, over hundreds of millions of years, sediments carried by rivers, sand grains blown by the wind, decaying plants and dead marine organisms were deposited in such basins. Sedimentary rocks formed, gas and oil were created, and were occasionally trapped in reservoirs. Geophysicists are employed by oil and gas companies to identify the sedimentary basins where oil and gas might have been formed and the structures where they might still be trapped.

On land, the identification of the right type of structure can sometimes be as simple a matter as observing the contours of the surface and looking for an obvious anticline. More often on land, and always where the geological clues are covered by the sea, the deeper structure of the rocks is concealed. This means that more scientific types of surveying are needed to give a general indication of the potential of an area.

The first rough picture of what a sedimentary basin might contain is provided by gravitational or magnetic surveys. Variations in the earth's gravitational pull can be measured by instruments on board a ship. As the ship travels along, the density of the rocks beneath the sea bed varies, affecting the gravitational field detected. It is possible to relate gravitational anomalies to the distribution of varying types of rock — for instance, an anticline can bring older and denser rocks nearer to the surface, increasing the gravitational pull. The same principles apply to magnetic surveys, using instruments in ships or aircraft to detect variations in the earth's magnetic field. As sedimentary rocks do not have significant magnetic properties, a magnetic survey indicates the depth

21

and form of the magnetic rock on which the sediments were originally deposited.

These surveys produce a broad picture of the extent of a sedimentary basin and the depth and distribution of sedimentary rocks. To obtain more details of what lies beneath the surface and to build up a map of geological structures, a more precise method must be used; the seismic survey. Seismology is the study of earthquakes, a study which makes use of the fact that the shock waves produced by earthquakes travel at different speeds through different rocks. When shock waves pass from one type of rock to another, they are reflected and refracted at the boundary, like light passing from air to water. The variations in shock waves make it possible for seismologists to make deductions about the rocks the waves have travelled through. In the same way, a seismic survey uses small artificial shocks to create waves which can be measured to give information about rocks beneath the surface.

To carry out a seismic survey at sea, two ships travel abreast along a straight line. One ship detonates a small explosive charge on the surface of the sea at fixed intervals, while the other tows a number of geophones, on a cable two or three kilometres long, which detect shock waves. The charge creates a shock wave which travels down through the sea and through the rocks beneath it, and is reflected and refracted as it passes from one layer to another. The geophones record the moment when the signal returns to the surface at each point along the length of the cable. These data are recorded, and later analysed to show the depth of the boundaries between layers of rock, and the types of rock involved. As the two ships move along a straight line, the seismic data make it possible to draw a continuous plot of the rises and falls of the rock strata. Further surveys build up a grid of observations, so that a contour map of each reflecting layer can be produced.

Seismic surveying like this is quick and relatively cheap, meaning oil and gas companies can acquire large quantities of data about the areas in which they are interested. To survey a 260 km² block of the North Sea would need a grid made up of 240 km of surveying. This would take five days and cost perhaps £150 per kilometre. In one year, one company alone acquired 30,000 km of seismic data, which perhaps indicates how indispensable this type of survey is in the early stages of exploration. The contour maps produced by analysis of seismic data show structures such as anticlines and salt domes far beneath the sea bed which might prove to be suitable traps where oil and gas have accumulated.

22

1. The jack-up drilling rig "Offshore Mercury", standing on its four legs, with the drill string supported from the derrick. The "Offshore Mercury" is here drilling an exploration well in the Irish Sea, where it discovered the Morecambe gas field.

Seismic surveying has rapidly become more sophisticated, with the use of non-explosive shock waves and much more complex analysis aided by advances in recording techniques and the use of computers. In some cases, it is even possible to detect the difference between a shock wave passing through sandstone saturated with water and one passing through a sandstone whose pores are occupied by gas. Even though information this precise can be obtained by seismic surveys, there is only one way to be really sure if there is gas or oil in what appear to be interesting structures — and that is to drill a hole.

Drilling an exploration well

Drilling a hole straight down into the ground is, in theory, a simple business. A cutting bit drills down into the rock, supported by a length of pipe which is in its turn suspended from a tower-like derrick and gradually lowered as the bit drills deeper. The pipe, and thus the bit, are rotated by the rotary table through which the

23

2. On the drilling floor of the "Offshore Mercury", the drilling crew attach another length of drill pipe to the string.

pipe passes. A drill pipe is 9 metres long, so each time that drilling has taken the bit 9 metres deeper, another pipe is suspended from the derrick and the lengths of pipe joined together to form a string. The drill string can eventually be as long as 3 km, built up of sections of pipe, with a bit at the bottom, all rotating. A string that long could weigh over 100 tonnes, so its weight has to be supported by a cable running over pulleys at the top of the derrick, thus regulating the weight allowed to bear on the bit. This weight determines the rate at which the bit penetrates the rock, and can be increased by drill collars, heavy sections of pipe added to the string immediately above the bit.

As the bit penetrates the rock, the hole it creates is cased — that is, the walls of the hole are lined with a steel tube and cemented. This prevents the hole collapsing, and leaves a space between the

24

rotating string and the steel casing of the walls. This space is used for the circulation of "mud" — a suspension of minerals such as barytes in water or oil, to form a thick fluid. The mud is pumped down through the hollow pipes making up the drill string, and returns to the surface along the space between the string and the casing. The composition of the mud is carefully controlled, as it performs several important functions: it cools and lubricates the bit, it helps to stabilise the walls of the hole and it carries fragments of rock back to the surface. If gas or oil is found, the weight of the mud forms a heavy seal in the hole, preventing gas or oil escaping to the surface under high pressure — known as a blow-out.

The mud is monitored as it returns to the surface. The rock fragments it carries show the type of rock through which the bit is drilling. The presence of hydrocarbons is indicated by dark spots in the mud or a film on the surface as it settles, evidence of oil, and by bubbles of gas in the mud. Samples of the rock itself are also taken for analysis, in the form of cylindrical cores cut by a hollow bit. Cores of up to 6 metres long may be taken every 15 metres when drilling in areas of special interest.

Eventually, using a lot of skill, good information from the geophysicists, and a measure of luck, examination of the mud and the core samples may show that the hole has been drilled in the right place, and that oil or gas has been struck. In this case, a production test is called for. A gun is lowered into the hole on a wire line to the depth of the interesting layer, and it is then fired to perforate the casing around the walls of the hole. Production tubing is lowered into the hole to a point just above the perforation, allowing oil, gas or water to flow to the surface under their own pressure. The production test goes on for several days, allowing the quality and quantity of the oil or gas to be examined and recorded. This information, together with the results of examination of core samples and measurements of the electrochemical and radioactive properties of the rocks made by instruments also lowered down the hole, provides a basis for the decision whether or not to carry out further investigations by more drilling, or to give up and try elsewhere.

Drilling rigs
This sequence of operations is indeed quite straightforward when carried out on land, as the ground provides the solid base on which the derrick and the drilling floor with its rotating table can rest. The same drilling process is used offshore, but in the middle of the

25

sea there is nothing solid to support all the equipment, plus the men who operate it. The stationary point from which drilling takes place has to be provided, and travel to the point where a hole is wanted.

The self-elevating drilling barge, more commonly known as the jack-up rig, does just this in the most obvious manner. It travels to its location like a ship, either under its own power or pulled by tugs. When it arrives, special legs are lowered which stand on the sea bed. When the legs are resting firmly on the bed, the whole vessel proceeds to jack itself up, so that it is well clear of the surface of the sea, standing on its legs. The limitation imposed by the length of these legs restricts the depth of water in which the jack-up can operate. This limit is generally taken to be a depth of 90 m, but where the sea bed is soft the legs may sink up to 9 m into the mud on the sea floor. There has to be a gap between the surface of the sea and the base of the rig when it is drilling, and in the North Sea in winter a gap of 15 to 20 m is necessary, if the hull is to be clear of the highest waves. Although this kind of rig can only operate in fairly shallow waters, these include the areas of the southern North Sea and Irish Sea where jack-ups found all the major gas fields. It has the advantage of placing all the drilling equipment on a fixed surface above the waves, while resting solidly on the sea bed.

In deeper waters, the drilling rig has to be supported not on legs but on the sea itself. There are two main types of floating drilling unit. The drill ship is simply a ship carrying a drilling rig, held in position with heavy anchors, while it floats on the surface. Waves and tides make it rise and fall, pitch and roll. The difficulty of drilling a hole at a fixed point on the sea bed from a point which is moving in all three dimensions has restricted the use of drill ships in the rough waters of the North Sea. These problems have been overcome to some extent, with the semi-submersible rig. This has large buoyancy tanks which are filled with air when it is travelling. On arrival at its location, the tanks are filled with water, so that the rig settles deeper into the sea. This increases its stability by reducing the effect of waves on the hull. The rig is kept in position by anchors and by propellers facing in all directions, which counteract the motion of the sea. As it is much more stable than a ship, a semi-submersible can work in water depths of hundreds of metres. This has made it very popular in the northern North Sea, where the Forties oil field, for example, is located under 125 m of water.

The presence of many metres of water between the drilling

floor on the rig and the point where it is boring a hole in the sea bed adds some complications to the work. With a stable jack-up rig, it is only necessary to drive a large hollow pipe from the floor of the rig into the sea bed, and then run the drill string through this drive pipe. The drive pipe keeps water out of the hole and permits drilling to be carried on from the rig floor, just as if it was on dry land, with all the well controls situated on the surface.

Drill ships and semi-submersibles pose more difficult problems, with their up and down motion linked to waves and tides. The connection between the rig and the sea bed through the water is made by a riser, with a telescopic section to compensate for movement. At the bottom of the riser there is a universal joint to allow lateral movement of the riser as the rig moves in the horizontal plane. The well head and the blow-out preventer are placed on the sea bed, at the base of the riser. Further flexibility is introduced into the drill string by the use of special sections of pipe, called bumper subs, above the drill collar. The rig rises and falls, but the bit must be kept in contact with the rock at the bottom of the hole, so that it continues to drill and is not damaged by bashing up and down on the rock. The collar's weight keeps the bit drilling at the bottom of the hole, while the bumper sub extends and contracts to absorb the motion of the rig.

Whatever type of drilling rig is used, it will of necessity be a large and cumbersome vessel. A crew of between 60 and 100 men will have to work, eat and sleep on it. The vessel has to carry not only the drilling equipment, the drill pipes and mud, but also cranes for lifting pipe and thousands of tonnes of supplies, as well as providing a pad where helicopters can land. The crew's work is always arduous, often uncomfortable and quite frequently dangerous. You have to be tough to do heavy manual labour in cold, wet and windy conditions, 12 hours a day, for two weeks at a time, followed by a week's leave on shore if you are lucky and provided the weather is good enough for a helicopter to lift you off the rig.

The high cost of building a rig — perhaps £15 million for a jack-up — and the high wages of a large crew of specialist workers are reflected in the price oil and gas companies pay to drill a hole. This can take at least a month, often much longer, and the price is constantly rising. In the northern North Sea, one exploration well can cost £3-4 million, and more often than not it will find nothing of any interest — a dry hole. It has been estimated that for each exploration well in the North Sea which found oil or gas in commercial quantities, there were 14 dry holes. Nevertheless, the

27

3. The Frigg gas field, in the northern North Sea. In the foreground, a drilling platform with a flare testing the gas. Behind it, a treatment platform and the quarters platform, linked by a bridge.

benefits of success are so great that they outweigh this ratio of success to failure, even at such high prices. By the late 1970s, 60 or 70 exploration wells a year were being drilled off the British coast.

The exploration well is the only way in which the conclusions of seismic surveys can be confirmed or disproved. On the rare occasions when oil and gas are found, then yet more drilling must be done, to find out how far the field extends and to appraise its potential for production. The appraisal of a field is carried out by drilling what are known as step-out or delineation wells. More holes in the same general area are drilled in the same way as exploration wells. From the evidence of the presence or absence of hydrocarbons, the decision can be made to develop the field. This is often a long and complicated process; after drilling one successful exploration well, which discovered the Beryl oil field, it was necessary to drill 18 more holes, nine of which were dry, to

28

delineate the field and establish the best method of developing it so that it could produce oil.

Developing a field

Exploration drilling establishes the presence of oil or gas, and appraisal drilling provides information about the potential for producing these fuels in commercial quantities. A further phase of drilling is necessary to develop the field for production. The relative importance of these three stages might be indicated by the number of wells drilled in British waters in one of the peak years of offshore activity, 1977. There were 67 exploration wells, most of which were of course dry, 38 appraisal wells and 96 development wells on fields found and delineated earlier by the first two types of drilling. There were also two exploration wells and three appraisal or development wells drilled on land in Britain that year.

Exploration and appraisal drilling is carried out by mobile rigs, which move from one hole to another. Development drilling takes place only when a commercial field has been identified, and is intended to create wells which will remain in use for many years, producing oil or gas. This drilling is therefore carried out from fixed structures, platforms designed specifically for each field, which stay in position until the field has been fully exploited.

If just one hole was drilled from a platform into a gas reservoir, and gas was allowed to flow up to the surface under its own pressure, water would gradually reoccupy the pores in the reservoir rock. In time, gas would have been drained from the area directly beneath the hole, and the incursion of water would prevent the flow of gas from the rest of the reservoir. When a field is found, it obviously makes sense to drain gas equally from all parts of the reservoir, so that as much gas as possible is produced before the field has to be abandoned. This means that a large number of holes have to be drilled into different parts of the reservoir.

Platforms are expensive devices, so it would obviously be wasteful to build a separate platform for each hole that has to be drilled, bearing in mind that a single field may need as many as 40 or 50 wells to drain it. The answer is to use the platform as the central point from which a large number of wells can be drilled downward and outward, to cover a wide area. This means directional drilling, as opposed to simple vertical holes, with each well deviated in the appropriate direction at the required angle, so that the platform is like the stump of a tree, with its roots under the ground extending for a considerable depth and distance all round it to gather water from the earth. In the same way, deviated wells gather oil or gas

29

which flows up to a central point.

The process of directional drilling, with the aim of eventually making a hole which extends up to 3 km below the platform and 1.5 km away from it in the horizontal plane, adds complications to the usual drilling techniques. It begins normally, drilling vertically downwards from the platform a few hundred metres into the sea bed. A point is then chosen where the hole can begin to bend away from the vertical, and this kick-off point is surveyed by lowering instruments on a wire inside the drill string. At the level of the drill collar, the instruments indicate the exact angle and compass direction of the hole.

To force the bit to drill in the desired direction, a steel wedge, called a whipstock, is placed at the bottom of the hole, so that the bit is turned slightly away from the vertical. Alternatively, instead of using the drilling table on the platform to rotate the whole drill string and the bit at the bottom of it, it is possible to use a turbine to drive only the bit. The mud pumped down the hollow pipes which make up the drill string powers the turbine, which rotates only the bit. A short piece of pipe bent to the correct angle is inserted into the string above the bit, and so the turbine-driven bit is forced to drill at an angle.

From time to time, the direction of the hole is checked, and altered if necessary, so that eventually it is angled between 15 and 30 degrees away from the vertical, with the angle building up at a rate of perhaps 3 degrees deviation for each 30 metres drilled. As it gets deeper, the hole is lined with a steel casing, cemented to its walls, and it is filled with the drilling mud. When the hole at last reaches the level of the gas-bearing rock, the casing is perforated by an explosive charge. The weight of the mud prevents the gas flowing until another string of hollow pipe, the production tubing, is run down inside the casing, providing the route for the gas to flow to the surface under its own pressure.

In the production tubing, about 100 m below the sea bed, there is a fail-safe valve which can stop the flow of gas. The valve is controlled from the platform, both manually and by automatic links to the fire-detection system, and operates to cut off the gas if there is any hazard. At the very top of the well, where the production tubing reaches the platform, there is a "Christmas tree" — a series of valves which looks a bit like a Christmas tree. These valves contain the pressure of the gas and regulate its flow. Once on the platform, the gas is treated, to remove most of the water and debris it has carried with it, and in fields where the natural gas pressure has fallen, the gas is compressed to improve its flow.

Each platform has many wells bringing gas to it for treatment — in the southern North Sea there are usually 24 wells from each platform. Once all the wells have been drilled, the drilling machinery is no longer needed, so it is dismantled, leaving just the platform and the gas production equipment. Normally, platforms are left unmanned, except for maintenance, remotely controlled either from a central platform or from the shore. Large fields need several platforms, each with its cluster of wells, to drain them adequately, particularly when there is a complex of reservoirs near to or above one another. Like all other aspects of offshore work, it is an expensive business. A simple platform, of the type used in the shallow waters of the southern North Sea, might cost £3 million to install, excluding the cost of the wells. The wells themselves cost about £750,000 each to drill. In the northern North Sea, the development of the oil fields in much deeper waters costs a lot more: the Beryl oil field required a platform, 40 wells and processing equipment amounting to a total of £120 million.

The need to build fixed platforms with their treatment plants and large numbers of wells obviously makes development much more expensive than exploration. In 1976, for example, all the exploration carried out in British waters cost £301 million. The total expenditure on development of oil and gas fields that year was £1,907 million, including £765 million for platforms, £624 million for equipment on the platforms and £134 million for wells. The costs of operating the fields which were already in production were only £134 million — a large sum by most standards, but considerably less than the initial investment required to find fields and install all the necessary facilities before the gas can be produced.

All the activity described so far only serves to find gas and bring it up from reservoirs far under the earth to platforms in the middle of the sea. Much still has to be done before the gas is of any practical use to consumers who want to heat their homes. Perhaps the best way of seeing how gas is brought from the field to the customer is to follow the process as it affects one particular field.

3

THE FRIGG FIELD –
FROM THE NORTH SEA
TO THE CUSTOMER

Exploration and discovery

The Frigg gas field is in the northern basin of the North Sea, about 360 km north-east of the Scottish mainland and 190 km west of the Norwegian coast. One of the largest gas fields in the North Sea, Frigg lies almost exactly in the middle of the sea, so that it falls partly into the sector belonging to Britain and partly into the Norwegian sector.

The North Sea is divided into sectors belonging to the coastal states because the minerals beneath the sea bed also belong to them, under international law. Each country therefore has the right to control exploration within its own sector, by granting licences to oil and gas companies. Britain and Norway have each divided their sector into blocks of about 260 km^2, and allocated licences for a number of blocks. These licences give the companies to which they are awarded the exclusive right to drill and produce on the blocks they cover, in return for payment of royalties on any oil or gas produced.

The story of the Frigg field effectively began when the Norwegian government offered a block, numbered 25/1, in the northern North Sea, on the western edge of the Norwegian sector. The block was known to lie in a potentially promising area, where geophysical surveys had detected a sedimentary structure suitable for the formation of oil or gas. The successful bid for the licence for this block was made by a consortium of French and Norwegian companies, who agreed to share the costs of exploration, with the risk of loss if it was unsuccessful and the chance of profit if oil or gas was found. The work of carrying out the exploration was entrusted to one of the members of the group, the French-owned oil company Elf Aquitaine.

The award of the licence for the Norwegian block in 1969 was

32

followed in 1970 by the award of the licence for the adjoining block in the British sector, designated 10/1, to another consortium, headed by another French-owned oil company, Total Oil Marine Ltd. At this stage, exploration of the northern North Sea was only just beginning. The first success was the discovery of the Ekofisk oil field in the Norwegian sector in September 1968, followed by the British Forties field in October 1970. As is always the case with exploration, Elf and Total were embarking on a risky venture, with no guarantee of success.

They began by carrying out a detailed seismic survey. Using advanced seismic techniques, they were able to detect not only a structure of the type which might serve as a reservoir, but also changes in the density of rocks under the sea bed which indicated the presence of hydrocarbons. Elf therefore drilled the first exploration well in block 25/1 in the Norwegian sector in June 1971, and were delighted to strike gas — although they might really have preferred to find oil. Appraisal wells drilled by Elf in the Norwegian sector and by Total in the British sector confirmed that they had found a large structure, extending into both blocks. The drilling was done by a semi-submersible rig, the P.81, anchored in water depths of nearly 100 metres.

The exploration and appraisal wells were drilled down for more than 1,800 m beneath the sea bed, until gas was found trapped in a layer of Eocene sands about 150 m thick, lying over about 2 m of oil. The gas was of good quality, with a composition of 95 per cent methane, 4 per cent ethane, a low sulphur content and small traces of the heavier hydrocarbon liquids. When the first exploration well was tested, it flowed gas at the rate of 700,000 m^3 a day, showing that the gas would flow easily and in large quantities under its own pressure. By the spring of 1972, the appraisal wells had demonstrated that the field contained more than enough gas to make it worth developing — about 230 billion m^3. The field was named Frigg after a goddess in Norse mythology.

Elf and Total decided to work together to develop the field, with Elf responsible for production of the gas and Total for its transportation to the shore. Negotiations began with British Gas about the sale of the gas, if it could be delivered on shore in Britain, and in 1973 a contract was signed to sell the gas from the British part of the field to British Gas. Another contract, for the gas from the Norwegian part, was approved by the Norwegian Parliament in June 1974. The whole field could thus be developed as a unit, with all the gas going to the same customer, although it was still necessary to settle the division of the gas between the two

33

countries involved. A treaty between Britain and Norway allocated 60 per cent of the gas to Norway and 40 per cent to Britain, in accordance with the way the field was divided by the boundary between the two sectors.

Developing Frigg

The North Sea above Frigg was 100 m deep, the deepest water in which any gas field had yet been developed. It was also very rough, with constant high swells, sudden storms and almost continuous wind and rain. Work was almost impossible in winter, so all the major tasks of construction for the field's development had to be fitted into the relatively calm months from May to September.

Elf soon discovered how difficult it would be to develop a field in the northern North Sea. The steel structure for the first drilling platform capsized as it was being towed into position in 1974 and was so seriously damaged that it had to be abandoned. The lost platform might have cost as much as £20 million; in addition, the whole development programme had to be put back while another platform was assembled.

The platforms for each field have to be specially designed and built, to deal with the particular requirements of water depth and the number of wells to be drilled. In the case of the Frigg field, Elf decided that two drilling platforms would be needed, each one drilling a cluster of 24 wells deviating out beneath it. Such a pattern would drain gas from the whole field with the maximum efficiency. Two further platforms would be required to treat the gas, removing unwanted liquids. A fifth platform would accommodate the crew and the control room, and a flare stack would be installed some distance away to burn off gas in an emergency.

Following the loss of the first drilling platform, the first season of work was the summer of 1975. At the peak of the construction activity, no less than 1,800 men were working simultaneously on a site in the middle of the North Sea. They used three barges, a fleet of supply ships, several helicopters to carry men and supplies to and from the shore and from one platform to another, a minisubmarine, diving support boats and a number of semi-submersibles anchored on site to serve as "hotels" for the work force. All the basic construction was completed and gas had begun to flow by September 1977.

One of the drilling platforms was built by the conventional means of installing an eight-legged steel frame, weighing 8,000 tonnes, and fixing it to the sea bed. The 800 tonne deck and 2,500

34

tonnes of drilling equipment were then assembled on top of the frame. The other drilling platform was built of concrete, which gave it the advantage that it could be partially fitted with equipment in sheltered waters near the shore, then floated into position. When it arrived on site, it was sunk to the sea bed, weighted with sand ballast and stabilized by its own weight — 150,000 tonnes of reinforced concrete. Each platform was used to drill 24 production wells. The gas arriving on the platforms from the wells is treated just to remove water and sand carried up from the reservoir, and to separate some liquid hydrocarbons, known as condensate.

The gas and condensate are sent in separate underwater pipelines from each drilling platform to its respective treatment platform, 650 and 700 metres away. The two treatment platforms are built of concrete, with treatment facilities resting on concrete columns stretching 133 metres from the sea bed to the deck. Treatment is chiefly a matter of drying the gas, removing more of its water content, and of injecting methanol, which prevents the formation of hydrate, an ice-like substance which can block pipelines. The treatment platforms also house pressure and flow controls, metering stations to measure the amount of gas produced, power stations providing all the electricity needed to operate the field and a central mud unit, from which mud can be pumped into the wells to stop their flow in an emergency. One treatment platform also provides room for the installation of compressors, which will be used to boost gas pressure when the natural pressure of the gas in the reservoir begins to fall. After treatment, the condensate is reinjected into the gas stream so that it can be carried to the shore together with the gas.

The treatment platforms are linked by a bridge 80 metres long, and another bridge connects them to the quarters platform. Walking from one platform to another high above the North Sea is a vertiginous experience, and little consolation is afforded to those with no head for heights by the knowledge that one is also walking from Britain to Norway. On the quarters platform there is sleeping accommodation for 120 people, the canteen, a library and a room used for films and recreation. Both men and women live and work on the platform, unlike the drilling rigs where women seem to be unwelcome even as visitors, so there is a semblance of normal community life in a small village perched on top of a steel tower in the middle of the North Sea. The quarters platform also houses the control room where the operation of the field is monitored, and the communications facilities, which keep the field in radio con-

35

tact with Britain and Norway. If necessary, all the operations on the drilling platforms can be remote-controlled from the quarters platform.

Each of the five platforms has a helicopter deck, as the helicopter is the normal means of transporting crew and light supplies from the shore to the field and from the quarters platform to the drilling platforms. Heavy supplies are transported by boat, but crossing the North Sea in a winter storm is an uncomfortable experience. Helicopters may be preferable, but bad weather may prevent them being used. Crew members are then unable to get off the platforms for shore leave.

Pipeline to the shore

The gas and condensate, after treatment, is still 360 km from the nearest point on the Scottish mainland. Total was entrusted with the job of bringing the gas ashore, which would mean passing through areas where the sea was over 150 metres deep. Total's task was as challenging as Elf's work on the platforms. When the field was fully operational, it would be producing 42 million m^3 a day, on average, at a pressure of 145 bar. To transport this quantity of gas, and the condensate produced with it, it was decided that two 800 mm diameter pipelines would be necessary.

The distance of the field from the shore, and the presence of condensate in the gas, added complications to Total's plans. It was also desirable to make allowances for the possibility that the pipelines might be needed to carry gas from other fields, as well as Frigg. Gas travelling along a pipeline gradually loses pressure, which reduces the quantity which can be carried. Under free flow, the capacity of each pipeline would be 30 million m^3 a day, which might not be as much as would be needed if other fields were to add their gas to the supply from Frigg. If the gas could be compressed, to increase its pressure again at a point between the field and the shore, then each pipeline's capacity would be increased to about 45 million m^3.

An intermediate platform, 186 km from the field and 174 km from the Scottish coast, would accommodate the compressors to boost gas pressure in the pipelines. A concrete structure 170 metres high was installed in 94 metres of water. The intermediate platform could also help to cope with the problems caused by condensate. Condensate, being composed of liquid hydrocarbons, tends to accumulate at low points in pipelines, forming "slugs" of liquid. In order to move the liquid along the pipeline, and to keep the inside of the pipe clean, plastic spheres known as "pigs" are

36

sent down the pipeline, pushed along by the pressure of the gas. Pigs are inserted on the treatment platforms and removed and reinserted at the intermediate platform.

The two pipelines were constructed from 60,000 lengths of pipe. Each piece of steel pipe was 12 metres long, with a diameter of 800 mm, walls 19 mm thick and the strength to withstand a pressure of 150 bar. In all, the pipelines involved 275,000 tonnes of steel. The pipe was manufactured in Japan and France, then transported to Scotland where each length was wrapped with bitumen to protect it against corrosion. A further coating of concrete was then added, which would protect the pipe and weigh it down so that it would rest securely on the sea bed.

The lengths of wrapped pipe were then shipped out to lay barges — vessels specially designed for offshore pipelaying. The process of pipelaying took place on a continuous production line principle. The lay barge was held in position with ten or 12 anchors, placed in front of and behind it by tugs. On the barge, lengths of pipe were welded together and fed out over the stern in a continuous line. As each section of pipe was fed out behind it, the barge moved forward the same distance, winching itself along its anchor cables. As the barge moved along, the welded pipe slid down a long half-submerged floating ramp, so that it sloped gently to its position on the sea bed.

A barge thus constructed a continuous stretch of pipeline, laid on the sea bed. The stretches of pipeline laid by different barges or in different seasons, as work had to stop in winter, then had to be joined together. Divers in air-tight chambers lowered to the sea bed welded the sections together, and connected the pipelines to the bases of the platforms. To protect the pipelines from damage, and to prevent them causing obstructions on the sea bed which might catch trawlers' nets, they were buried. A sled pulled by a barge travelled along the pipeline, cutting a trench beneath it with high-pressure jets of air and water. The pipe settled into the trench, and was buried as currents filled in the space.

Three lay barges, two mini-submarines to check on the progress of pipelaying and trenching, and more than 60 other ships carrying supplies, pipe, divers and equipment were needed to lay the two pipelines. The pipelaying programme took Total more than three years, but when the pipelines were completed and connected to the platforms, gas could begin to flow ashore.

The shore terminal
The twin pipelines carrying natural gas and condensate from the

4. *The St Fergus terminal on the Scottish coast, where gas from the Frigg field comes ashore, showing some of the pipework and four of the compressor cabs.*

Frigg field reach the Scottish coast at St Fergus, 64 km north of Aberdeen. Here at the terminal, Total delivers the gas to British Gas, as agreed in the contract which specifies the quantity and quality of the gas to be purchased. The contract lays down the acceptable limits for temperature, pressure, calorific value and proportions of water, sulphur and liquids in the gas. Much of the activity on Total's part of the terminal is devoted to ensuring that these specifications are achieved before the gas is passed to the British Gas terminal for transmission to gas consumers.

The first stage of the treatment of the gas and the slugs of condensate, pushed along by pigs, which arrive with it, is to remove the condensate. British Gas will not accept liquids in the gas it buys, as it would tend to collect in pipelines, particularly if condensation took place when pressures were reduced. On the other hand, liquid hydrocarbons are very useful as chemical feedstock. When the two streams of gas come ashore in the pipelines, the first thing to be done is to remove the pigs as they arrive. The gas then

38

passes through a "slug catcher". At St Fergus, there are eight parallel runs of pipe 207 metres long, with a gentle slope. The slugs of liquid are deposited at the lowest point of this slug catcher, while the gas flows on. As a further measure to remove traces of liquid hydrocarbons or water, the gas then passes through refrigeration units, lowering its temperature to -18°C. At this temperature, nearly all the liquids are condensed out of the gas. The condensate which has been separated from the gas is stored in tanks, and taken away from the terminal by road tankers, for use in the chemical industry.

After tests of its quality and metering of its quantity, the dry gas then passes from the Total terminal to the British Gas terminal. Further checks confirm exactly how much gas British Gas is receiving, and an odorant is added to it. When it is produced from the field, natural gas has little or no smell, so an artificial odour has to be added as a safety measure. The smell, which has to be distinctive and pungent so that gas leaks can easily be detected, is provided by a mixture of chemicals with an odour so strong that only 1 kg of odorant is needed for 60,000 m³ of gas.

British Gas receives gas from the producers at St Fergus at a pressure of between 40 and 50 bar. This pressure is below the operating pressure of gas transmission pipelines in Britain, 69 bar, and it is obvious that the higher the pressure of the gas, the larger the quantity which can be sent through a pipeline of a given diameter. At St Fergus, British Gas has installed eight compressors to raise the pressure of the gas, with a total capacity sufficient to handle the compression of 100 million cubic metres a day. This allows a surplus of compression capacity to handle additional supplies of gas from fields other than Frigg, and to permit one or more compressors to be taken out of service for maintenance, without hindering the continuity of supply. These compressors have a maximum compression ratio of 1.62:1. As anyone who has used a bicycle pump can testify, the compression of a gas raises its temperature, in this case from 25 to 85°C. Such a high tempeature would damage the protective wrapping on pipelines, so after-coolers reduce the temperature of the gas to an acceptable level, below 50°C.

Most of the plant at St Fergus is needed simply to handle the gas, to treat, dry, meter, odorize, compress and cool it, so that it is ready to pass along the pipelines which will eventually bring it to the consumers. The gas arrives in two streams, and all the plant on the terminal is duplicated, at least, so that if any part of the process breaks down or needs to be serviced, gas can still flow

39

through the unaffected plant. If, for some reason, gas arrives on shore which cannot be passed through the terminal, it can be flared via vent stacks, which are simply chimneys 20 metres high, standing in isolated areas.

In addition to the plant handling the gas, many other facilities are needed on the terminal. Control rooms monitor the flow of gas. Continuous communications have to be maintained with the platforms offshore and with the central control of the onshore gas transmission system, so that the flow of gas can be related to the fluctuations of supply from other sources and demand from consumers. Standby electricity generators provide power for the terminal if the normal electricity supply breaks down. An artificial lake provides a water supply for fire-fighting in an emergency. In all, the producers' terminal and the British Gas terminal each occupy nearly 50 ha.

The whole terminal cost about £100 million to construct, while the cost of the platforms on the Frigg field and the offshore pipelines was about £1,800 million. This investment was needed to handle 42 million m^3 a day of natural gas, the peak flow from Frigg, about one-third of Britain's total gas supply in the late 1970s. Frigg is only one of several gas fields, and St Fergus one of four North Sea shore terminals, supplying Britain with gas, as will be seen in the next chapter; but bringing the gas from the field to the shore is only a part of the story.

Pipelines on land

Gas from Frigg has to be fed into the national gas transmission system, which carries natural gas from the North Sea to all parts of Great Britain. British Gas has built an interconnected system of high-pressure pipelines, which transport large quantities of gas from the terminals to the twelve regions of British Gas, where it is distributed to consumers. The transmission system, consisting of pipelines buried under the ground, transports energy silently and invisibly, without causing environmental pollution or additional traffic on overcrowded roads. Investment in the transmission system has amounted to over £2,000 million.

The natural gas transmission system began as a single 600 mm diameter pipeline, built in 1964 to carry the Algerian natural gas which was then first unloaded at Canvey Island in the Thames estuary. This pipeline, constructed before natural gas was found in the North Sea, ran 360 km from Canvey Island to Leeds, with short spur lines to London, Reading, Birmingham and Manchester. The first North Sea gas field was discovered in 1965, and gas began
40

to come ashore in 1967. As more fields were found, more shore terminals were built, and pipelines to carry the gas were laid, joining together so that the whole of Britain was eventually served by a single integrated system. Most of the pipelines are 900 mm in diameter, and their total length now exceeds 5,000 km.

When the Frigg field was discovered, and the gas was bought by British Gas, for delivery at St Fergus, the transmission system extended only as far as the south of Scotland. A single 600 mm pipeline brought gas from the southern North Sea fields through northeast England to a point between Edinburgh and Glasgow. Frigg would add a large additional quantity of gas to the available supplies from the southern North Sea, so St Fergus would have to be linked to the existing pipelines, and the system extended so that gas could flow south into north-west England. To handle Frigg gas, initially two pipelines were laid, running south through Scotland from St Fergus, then on into England. Construction of 950 km of 900 mm pipeline began in 1974, and was completed in 1976, at a cost of £140 million.

The first stage in the project was to find a route for the pipelines which would avoid both built-up areas and, as far as possible, the most difficult terrain. In northern Scotland, the former requirement is more easily fulfilled than the latter, and the route had to cross peat bogs and mountainous sections, as well as 486 rivers and streams. Once a possible route had been worked out, permission to lay the pipeline had to be obtained by negotiation with landowners and local authorities. This required a lengthy series of consultations, with British Gas offering assurances of minimal disruption and promises of financial compensation, until farmers and councillors were persuaded to agree on the necessity of a pipeline, and to accept that it could be constructed without causing permanent damage to their land and their pockets.

High quality steel pipes, with walls 12.7 mm thick, were ordered from several manufacturers, tested for soundness, then wrapped with coal tar enamel and fibre-glass to protect them from corrosion. The contractors who actually laid the pipes marked out a corridor along the length of the route, then dug a trench deep enough to allow at least a metre of cover on top of the pipe once it had been laid. Top soil was placed to one side of the trench, subsoil to the other, so that they would not be mixed. The pipes were then delivered to the site, and welded together in sections beside the trench. Each weld was X-rayed to ensure that the joints were sound, and the pipe electrostatically tested to detect any defects. The welded sections of pipe were then lifted by a row of

41

5. Pipelaying before . . .
42

cranes and placed in the trench. After the pipes had been joined, the trench could be filled, replacing first the subsoil, then the top soil. The pipeline was also pressure-tested, by filling it with water to ensure that it could withstand a pressure much greater than it would be subjected to when in use.

Special techniques had to be used for some of the more difficult parts of the pipeline. It crossed 576 roads and 23 railway lines, and clearly major roads or railways above the pipe would subject it to exceptional stresses. To protect the pipe, it was taken through steel sleeves 1050 mm in diameter. The ends of each sleeve were sealed to the pipeline, and the space between them filled with nitrogen, so that, if the sleeve was not enough to protect the pipe, any escape of gas would still be contained, and not mix with air. These roads and railways posed yet another problem, as traffic could not be stopped for weeks on end while a trench was dug across them and the pipe laid. The pipe was therefore taken through a hole bored beneath them, rather than being laid in a trench. The same technique was used when the pipeline crossed Hadrian's Wall, an archaeological feature far too important to be disturbed.

Laying a pipeline inevitably causes considerable disruption while it is taking place, but the trench, the digging machines and the cranes are all confined within a narrow corridor. When construction has been completed, and the soil put back in place, the land returns to its original state remarkably quickly. The overall aim in laying the pipeline was to leave as little trace as possible of its passage. In just a few years, the reinstatement of the land was virtually complete, with crops growing and animals grazing above it, just as they were before the pipeline arrived. The only restriction was that trees should not be planted too close to the pipeline, in case their roots damaged the pipe-wrapping. Walls and fences were rebuilt to match adjoining sections which had not been taken down, and farmers were paid compensation for the loss of the use of their land and of their crops during construction.

After a few years, the only trace left of the pipeline was the line of marker posts placed along it so that the route could be followed from the air. Helicopters fly along the whole length of the system once a fortnight, to keep an eye on excavation or other works which might affect the buried pipes. In addition, the whole length is regularly surveyed on foot, and contact is maintained with farmers and landowners, to avoid any accidental damage to the pipes.

The first two pipelines from St Fergus were used to carry gas

43

6. . . . *and after. Only a year after a high pressure gas pipeline was laid, the only trace of its passage is the marker post which allows helicopter pilots to follow the route of the pipeline.*

from the southern North Sea northwards into Scotland, and, when gas from Frigg began to flow in 1977, to transport it into the national system. The prospect of further supplies arriving at St

44

Fergus from other fields led to the construction of a third 900 mm pipeline, completed in 1978. A fourth pipeline through Scotland was also planned, as it became plain that still more gas could be expected from the northern North Sea. This pipeline, to be constructed between 1980 and 1982, is to be constructed from 1050 mm diameter pipe, to increase its capacity.

The national transmission system is still growing, but it has been laid in much the same way as the lines from St Fergus, to carry gas at high pressure unseen and unheard, in pipelines that leave no trace behind them to scar the countryside. The safety of the pipelines is assured by the use of the strongest materials, thoroughly examined and tested before they are brought into use. Should there ever be a gas escape, caused for instance by the pipe being pierced by an excavator, the resulting drop in pressure would be detected immediately, and remotely operated valves in the pipe would slam shut. The affected stretch of pipe would thus be isolated, minimizing the loss of gas, and remain out of service until it was repaired. As more pipelines are added to the system, it becomes less vulnerable to such accidents, as alternative pipelines are available to make up for the loss of a damaged one. However, this is merely a hypothetical consideration. Throughout the history of the high-pressure gas transmission system in Britain, there has never been a single break, or leak, or any other sort of accident.

Of course, this perfect record can partly be attributed to the fact that most of the system is quite new, and was laid only after thorough testing. In time, all pipes and welds are subject to deterioration. The detection of flaws in a buried pipeline is obviously difficult, particularly if one considers the problems of uncovering a pipeline and taking it out of service just to make sure that it is in good condition. British Gas has developed a device which seems to provide the answer, the "intelligent pig". Like ordinary pipeline pigs, this travels inside a pipe, pushed along by the pressure of the gas. This particular pig is equipped with sensors which can detect any flaws in the wall of the pipe caused by corrosion or external damage, and records exactly where they are. The transmission system pipelines have been equipped with points where pigs can be put into the pipes and taken out, at regular intervals. The intelligent pig can thus be put into a stretch of pipeline, travel along it with the gas, recording the condition of the pipe, and on removal from the pipe, allow its state to be examined, and flaws pinpointed, without taking the pipe out of service. British Gas is developing a series of pigs, of various diameters to fit the range of pipe sizes, which will help to ensure the safety of pipelines as they

45

get older.

Compression

Gas leaves the shore terminals and enters the national transmission system at a maximum pressure of 69 bar. As it travels along the pipelines, the pressure drops, reducing the quantity of gas which the pipelines can carry. To move more gas along the pipelines, compressors are used to boost its pressure back up to the normal operating level, just as they are on the offshore platforms and at the St Fergus terminal.

The optimum way of operating pipelines as efficiently as possible is to build compressor stations every 65 km along the pipeline, to raise the pressure by a ratio of 1.4:1. This achieves the best balance between keeping up the pressure and transmitting the greatest amount of gas, on one hand, and paying for a compressor station and the running costs on the other. A compression ratio of 1.4:1 is favoured for compressor stations on the transmission system because it does not raise the temperature of the gas to such an extent that aftercoolers are necessary as is the case at St Fergus, where a higher compression ratio is needed. A pipeline with a diameter of 900 mm, the most common size in Britain, can transmit 35 million m³ a day with compressors every 65 km. If the compressors are spaced out more distantly, say 130 km apart, the capacity of the pipeline falls considerably, to 25 million m³ a day.

At the moment, much of the transmission system does not have to be used to its full capacity, having been constructed with sufficient surplus capacity to cope with the gradual increase in gas supplies in the 1980s, as more gas becomes available from the northern North Sea. Although sites have been outlined for compressor stations at the optimum distance points, they are still more sparsely distributed. As gas consumption rises, more compressors will be built. The three 900 mm pipelines running through Scotland from St Fergus, for example, are expected initially to carry an average of 42 million m³ a day from Frigg. In the 1980s, 18 million m³ a day will be added to this by the Brent field, as well as further quantities from the other northern fields. These are only average figures; gas demand on a cold winter day can be 50 per cent higher than on an average day, so clearly pipelines have to be able to carry larger than average supplies in winter. Nevertheless, the three Scottish pipelines can transmit 75 million m³ a day with compressor stations 130 km apart. As supplies build up, compressor stations can be added at the intermediate 65 km spacings, increasing their capacity to 105 million m³ a day.

46

7. A compressor station, with two cabs housing the compressor machinery and silencing the noise of its Rolls-Royce Avon engines.

Compressor stations thus play a vital role in keeping large quantities of gas flowing through the system. At each station on the Scottish pipelines, gas is compressed from 48 bar at 5°C to a maximum of 65 bar at 45°C. The power to drive the compressors at the stations on the transmission system and at St Fergus is provided by industrial versions of Rolls-Royce jet engines. The engines are fuelled by natural gas from the pipeline, and the hot gases they produce drive gas turbines, which in their turn drive the compressors. At St Fergus there is an array of eight compressors, while each compressor station has two or more, depending on the amount of gas it has to compress, and the allowance which has to be made for standby compressors in case one breaks down.

The engines generally used in compressor stations are the Rolls-Royce Avon and Maxi-Avon, which are better known as the engines used in the Canberra, Lightning and Hunter military aircraft and the Comet IV civil aircraft. The larger RB211 is the engine Rolls-Royce developed for the Lockheed Tristar airliner, which is to be used in compressor stations where more power is required to handle larger quantities of gas. The noise produced by such aero-engines is only too familiar to anyone living near an airport. At compressor stations, the air intake and exhaust are silenced, and the whole compressor unit housed inside an acoustically insulated cab. Inside the cab, the noise level of 120 decibels would be unbearable without some form of protection for the ears. 100 metres away, so effective is the insulation, the noise level is only 40 decibels. The criterion used in designing compressor stations is that their operations should be inaudible at the nearest house at the quietest time during the night.

Compressor stations tend to cover fairly large areas, perhaps 15 to 25 ha for some of the major ones which compress the gas from two or three pipelines running parallel. Much of this space is occupied by buried pipework, so that little is visible, apart from the acoustic cabs housing the actual compression plant, and a small building for offices and communication equipment. The staff at each compressor station need number no more than ten or twelve; in theory, stations could be left unmanned, and controlled remotely, though this is not done in Britain.

Distribution

The pipelines of the national transmission system carry gas from shore terminals like St Fergus to all parts of Great Britain, with compressor stations maintaining the high pressure necessary to transmit large quantities of gas over long distances. The transmission system is a unified, integrated network, so that the whole country is linked together in a single grid which can carry gas from any of the terminals to any region of Britain. The regions of British Gas, which supply gas to the consumer, receive gas from the transmission pipelines at more than 120 offtakes.

At these offtakes, the pressure of the gas is reduced from the transmission pressure of 69 bar, and the gas fed into medium-pressure pipeline systems operated by each gas region. These pipelines carry gas to centres of demand, such as major towns, at pressures below 38 bar. The pressure is then reduced again, to between 2 and 7 bar, for the distribution of the gas through more localized networks of pipes. Finally, the familiar gas mains which run beneath the streets of every town are filled with gas at pressures of some 25-35 mbar, and service pipes take the gas from the mains to the meter in each customer's home.

The whole gas supply system might be compared with the road transport system, in broad terms. The high-pressure pipelines of

48

the national transmission system are like motorways, carrying large quantities over long distances. As traffic leaves the motorways for A-roads, so gas passes through the offtakes for the regional medium-pressure networks. The low-pressure system might be likened to B-roads, and street mains to the small streets in towns, beneath which they run. To complete the analogy, the service pipe is like the path running from the street to the house.

The gas system, like our road system, was not designed as a whole, but grew haphazardly, with the unifying force of the transmission system being imposed on regional and local grids which were already in existence. Much of the distribution system goes back to the days when gas was manufactured from coal at gas works in each town. Each works had its local system of gas mains radiating out from it, running beneath the streets. Some works were linked by medium-pressure pipelines, particularly in the 1950s, in an attempt to concentrate gas manufacture on the most efficient works and develop regional grids, rather than having isolated works in each town. When natural gas was discovered, the national transmission system was constructed, and each region linked its grid to it.

One of the disadvantages of the random way in which the gas supply system developed over nearly 150 years is the fact that the street mains were intended to carry town gas, made from coal, rather than natural gas. The high-pressure pipelines were designed specifically for natural gas, but although they carry gas in bulk, they make up only a tiny proportion of the total length of gas pipes in Britain. In all, the gas distribution system is made up of some 220,000 km of mains, together with 14 million service pipes.

These mains and service pipes were originally constructed of cast iron, for the most part, and the joints between lengths of pipe were sealed with lead and jute fibre. Town gas made from coal contained liquids which kept the fibre moist and swollen, thus sealing joints. Town gas made from oil was drier, and natural gas is drier still. When the fibre dried out, it no longer formed an effective seal, and leakage from joints became a major problem. This leakage has been contained by injecting glycol as a mist into the gas carried in vulnerable mains. The humidified gas keeps joints moist and swollen, preventing leaks.

The use of cast iron poses more difficult problems. In some soils, it is subject to corrosion, although in others it can last almost indefinitely. In addition, cast iron is liable to fracture when placed under stress. Mains laid many years ago beneath roads and pavements are now subject to stresses which could not have been envis-

49

aged then, from heavy lorries and vehicles parked on the pavement. Leaking mains can be repaired, but the long-term solution to leakage from corroded mains, or mains particularly liable to fracture, is their replacement. To replace the whole low-pressure gas distribution system would be impossibly expensive, as well as unnecessary — it could cost £4,000 million. Most mains are in good condition, and much of the system has been laid relatively recently, but those parts of the system which are most vulnerable, or have a record of leakage, are replaced as the opportunity arises. Over the ten years from 1968 to 1978, some 36,000 km of new mains were laid, and 7,800 km of old mains were relaid. The programme of mains replacement was accelerated in the late 1970s, with a planned investment of £500 million intended to achieve the replacement of those mains most at risk by 1984. Similarly, the expansion of gas supply led to the laying of 3 million new services between 1968 and 1978, and the relaying of 2.5 million old services, so that a large proportion of the services bringing gas from the mains into the house are quite recently installed.

For the most part, mains and services are now made of polyethylene, rather than cast iron. Polyethylene is flexible, which means that it is less liable to fracture when subject to the stresses produced by heavy traffic and earth movement. The lightness and ease of jointing of polyethylene make it much easier to work with than cast iron, and its flexibility has the advantage that in the replacement of old mains a polyethylene pipe can simply be inserted inside the old main, thus saving the effort of digging a trench and causing disruption of traffic.

As the number of holes in the road makes clear, the repair, maintenance and replacement of distribution mains is a continuous and widespread process. It is at the level of the local distribution mains that the modern natural gas industry encounters the relics of the old town gas industry. Natural gas, produced from the Frigg field using the most advanced technology, eventually reaches the consumer through mains which might well have been laid by some long-vanished gas undertaking in the days when gas was made from coal, and the only thing produced from the North Sea was fish.

4

SUPPLY AND DEMAND

Fields and terminals

The story of the discovery and development of the Frigg field, and the way in which gas is brought from the field to the consumer, might serve as an example of the principles of producing natural gas from the North Sea, but Frigg is far from being the only field supplying Britain. Frigg only started production in 1977; for ten years before then, gas had been arriving from fields in the southern basin of the North Sea, off the east coast of England.

The southern fields are closer to the shore and in shallower waters than Frigg. As a result, although they use the same types of platforms for drilling production wells in clusters and treating the gas, and pipelines to bring it ashore, the platforms can be smaller and the pipelines shorter. The average water depth in the southern North Sea is only about 25 m, so platforms are about 60 m high, from the sea bed to the top deck. The pipelines to the shore terminals, with diameters ranging from 400 to 750 mm, are from 30 to 80 km long.

As with Frigg, the southern North Sea fields are operated by the companies which discovered them, and the producing companies deliver gas to shore terminals, where it is bought by British Gas for supply to gas consumers within Great Britain. The shore terminals resemble St Fergus, in that the gas is treated to ensure it complies with the specified standards of dryness and quality, then passed to British Gas for the addition of odorant and transmission by pipeline. At St Fergus, gas pressure has to be increased to the transmission pipeline pressure of 69 bar. At the other terminals, the gas arrives at higher pressures, which may have to be reduced. Instead of aftercoolers, these terminals need to heat the gas to prevent the low temperatures produced by a sudden pressure drop causing ice formation. That apart, the terminals for the southern North Sea

51

fields perform much the same functions as St Fergus.

The first gas field to be discovered in the southern North Sea was West Sole, found by BP in 1965. A terminal was built at Easington, on the Yorkshire coast, and gas first arrived there in 1967. Most of the gas produced from West Sole is sold to British Gas, but BP takes a small proportion of the gas direct from the terminal to a chemical plant, for use as feedstock. Easington also serves as the terminal for gas from the Rough field, which began production in 1975.

The second terminal to be built, and still the largest, is at Bacton, in Norfolk. It began to operate in 1968, receiving gas from the Leman Bank field, and now also handles gas from the Hewett and Indefatigable fields. Each of these fields is operated by a different group of companies, so there are three producers' terminals at Bacton, each treating gas before passing it to the British Gas terminal. In all, Bacton could handle a maximum throughput of 112 million m^3 a day from all the fields — energy equivalent to a 1,000 tonne trainload of coal every 10 minutes all day long.

The Theddlethorpe terminal, on the Lincolnshire coast, takes gas from just one field, Viking, which began production in 1972. Theddlethorpe does, however, make provision for extensions which might be necessary if the terminal were required to cope with additional supplies from other fields.

We have already seen how St Fergus handles gas from Frigg, and how the pipelines from Frigg and the intermediate platform were built with sufficient capacity to handle additional supplies from other fields in the north. The St Fergus terminal itself was designed in the knowledge that Frigg would be only the first source of gas from the north, and that another major field, Brent, would also send gas there. The producers' terminal for Brent gas, next to the Total terminal for Frigg, should start operating in 1980/81. At present, the Frigg pipelines are already carrying small quantities of gas from Piper, a field near the intermediate platform, to St Fergus.

There are two other sources of natural gas for Britain, which do not involve the four North Sea gas terminals, both of relatively minor importance. The Forties oil field is linked to the Scottish coast by a pipeline which comes ashore at Cruden Bay, carrying some gas dissolved in the oil. A land pipeline carries the oil and gas to a refinery at Grangemouth, near Edinburgh, where the gas is separated from the oil, and sold to British Gas.

The other minor source of natural gas does not involve the North Sea at all. Liquefied natural gas (LNG) from Algeria is brought by ship to a terminal on Canvey Island, in the Thames estuary. This terminal opened in 1964, before gas was discovered in the North Sea, and has now become relatively unimportant. Since 1964, two LNG carriers have shuttled back and forth between the terminal at Arzew on the Algerian coast and Canvey Island, so that one arrives at Canvey every five days or so. This supply is equivalent to less than 3 million m^3 a day, but is still quite useful because it provides LNG which can be stored at Canvey and used to top up supplies to the London area when demand is particularly high.

The role of Algerian LNG relative to North Sea gas can be gauged by the fact that total gas supplies in 1977 were equivalent to 124 million m^3 a day. Nearly all this gas, 108 million m^3 a day, came from the fields in the southern North Sea. These daily figures are averages; obviously, gas demand is much higher in winter than in summer. To put it another way, in the course of the whole year of 1977, Britain received 41.79 billion m^3 of natural gas — 39.96 billion m^3 from fields in the British sector of the North Sea, 952 million m^3 from the Norwegian part of the Frigg field and 877 million m^3 from Algeria. Of this gas, 163 million m^3 was gas from West Sole supplied direct to a chemical works, leaving over 40 billion m^3 of natural gas supplied to British Gas. (See Table 1).

Gas supplies in the 1980s

In 1977, most of the gas arriving in Britain was gas from the southern North Sea, delivered to the three terminals on the east coast of England. Production from Frigg was only just beginning, and from Forties was negligible. Frigg began production in September 1977, and by the end of the year had totalled 1,565 million m^3 of gas from the British and Norwegian sectors together. The production rate of about 3.9 million m^3 a day from Frigg at the end of 1977 would soon be left behind, as gas flow rose towards its maximum rate of 42 million m^3 a day, expected in 1980.

The increase in gas supply from Frigg is just the first in a number of additions to Britain's gas supplies expected in the 1980s. The fields in the southern North Sea will then be at or past their peak production, but more gas from the north will be following Frigg. Unlike Frigg and the southern fields, the new sources of gas in the north will be primarily oil fields. Many oil fields contain associated gas, which in the past, in the traditional oil-producing countries of the Middle East and the USA, was flared. In other words, it was thought to be of little value, so it was disposed of by

53

Britain's Offshore Natural Gas Supplies

FIELD	LICENSEES	DATE DISCOVERED	PRODUCTION STARTED	GAS BROUGHT ASHORE 1977 (million m³)	RECOVERABLE RESERVES (billion m³)
West Sole	BP	Oct. 1965	Mar. 1967	1,944	61
Leman Bank	Shell/Esso	Apr. 1966	Aug. 1968	15,525	292
	Amoco/British Gas				
	Arpet group				
	Mobil				
Hewett	Arpet group	Oct. 1966	July 1969	7,776	97
	Phillips group				
Indefatigable	Amoco/British Gas	June 1966	Oct. 1971	6,760	125
	Shell/Esso				
Viking	Conoco/BNOC	May 1968	July 1972	6,285	125
Rough	Amoco/British Gas	May 1968	Oct. 1975	1,057	14
Frigg (UK)	Elf/Total	May 1972	Sept. 1977	613	128
Frigg (Norway)	Elf/Total	June 1971	Sept. 1977	952	192
Forties	BP	Oct. 1970	Sept. 1977	—	n.a.
Piper	Occidental group	Jan. 1973	Nov. 1978	—	n.a.
Brent	Shell/Esso	July 1971	1980/81	—	84
Tartan	Texaco	Dec. 1974	1980/81	—	n.a.
Morecambe	British Gas	Sept. 1974	1984	—	56-84

Note: Estimates of reserves for Forties, Piper and Tartan gas are not available, as production depends on oil recovery.

burning it. Today, it is realized that natural gas is too valuable to waste like this, and the British government has taken a very firm lin with the oil producing companies in the North Sea to prevent flaring. The general principle is that development should only be allowed if adequate steps have been taken to make good use of any gas produced with the oil. This might mean using the gas as a source of power on the production platforms, or reinjecting gas back into the reservoir to aid oil production, but it can also mean finding a way to bring the gas ashore.

The delivery of gas from Forties to the shore, together with oil, was the first example of this policy of recovering associated gas. It was followed in 1978 by the construction of a short pipeline to take gas from the Piper field to the intermediate platform on the Frigg pipelines, so that it could be delivered to St Fergus with Frigg gas. In the early 1980s, Forties will be producing a mere 0.25-0.5 million m^3 a day, and Piper will have reached its peak rate of perhaps 1.8 million m^3 a day. These may seem unimpressive quantities, but they are just the first stages in the process of recovering a vast quantity of associated gas — it has been suggested that as much as 300 billion m^3 of gas and condensate could be produced from the oil fields of the northern North Sea.

The first major source of associated gas will be the Brent field, which is due to start production in 1980/81. This field is so large that, as well as oil, it will also produce 18 million m^3 a day of gas at its peak, enough to merit its own pipeline to the shore and another producers' terminal at St Fergus. The large quantities of gas from Brent and Frigg, together with some associated gas from other northern fields, should increase Britain's gas supplies from about 125 million m^3 a day in the late 1970s, to 160 million m^3 a day in the late 1980s. Further supplies of associated gas should enable this rate to be maintained, despite the decline of the southern fields, into the 1990s.

It seems likely that most of the associated gas from the northern oil fields will be brought ashore like the gas from Piper, by constructing pipelines which will link the fields with the major pipelines from Frigg and Brent. Piper itself is to be linked to the Claymore and Tartan fields. Two extensions to the Brent pipeline are already planned, one running north to the Magnus field, with links to Murchison and Thistle, the other running west to Cormorant, and also taking gas from Heather, Ninian and North West Hutton.

These gas collecting pipelines are only the first to be planned. Many more oil fields with associated gas are already known in the

55

northern North Sea, and more may be discovered. The most probable pattern of development seems to be that oil and gas should be separated on the platforms, and where there is surplus gas, it should be sent ashore by pipelines linked to the Frigg and Brent trunk lines. The whole picture could be changed if Norway decides to develop some of the fields in the north. One field, Statfjord, contains considerable quantities of gas as well as oil, and, like Frigg, it lies partly in the British sector. It is possible that gas from Statfjord and other Norwegian fields could be sold to British Gas and brought ashore in Britain, but as yet no decision has been reached.

Whatever happens to the associated gas in the northern North Sea, it will presumably be brought ashore at St Fergus, which has the capacity to handle very large quantities of additional gas. The pipelines linking St Fergus to the rest of the national gas transmission system are to be joined by a fourth pipeline running through Scotland, to be constructed by 1982. With this additional capacity, further supplies of gas can be introduced into Britain in the 1980s.

Also in the 1980s, a new terminal will be built, not on the shores of the North Sea, but on the other side of Britain, on the Irish Sea coast. British Gas has discovered a major gas field, the Morecambe field, some 30 km west of Blackpool. Reserves are estimated at between 56 and 84 billion cubic metres, following several appraisal wells, and it is planned to build a terminal in the north-west to receive the gas. Production from the Morecambe field could begin in 1984, if all goes well, but development plans are as yet not complete.

The development of the Morecambe field and the recovery of the associated gas from the northern North Sea are the major additional sources of gas foreseen for the 1980s. Together with the increasing production from Frigg, they should permit a level of gas supply some 50 per cent greater than was the case when all our natural gas came from the southern North Sea. What happens after that is, in the nature of things, much more difficult to predict.

Gas reserves

Nobody knows how much gas there is under the North Sea, or elsewhere around Britain. After all, the first gas discovery in the southern North Sea was as recent as 1965, while Frigg, in the northern North Sea, was found in 1971. There is much more exploration still to be done in the North Sea, while in other parts of Britain's continental shelf exploration has hardly begun. The discovery of the Morecambe field in the Irish Sea has demonstrated that gas is not only to be found in the North Sea, but of course only exploration drilling can demonstrate whether more gas remains to be found.

Drilling has only just begun west of the Shetlands, but oil has already been found there. In Cardigan Bay and the Celtic Sea, off the coast of Wales, drilling has found little of interest, although the Kinsale Head gas field has been discovered in the Irish sector of the Celtic Sea. The area known as the Western Approaches, beyond the Scilly Isles, has remained almost untouched so far, and the first exploration well in the English Channel was drilled in December 1978. There is evidently a lot of exploration still to be done, and the rapidly advancing technology of offshore drilling is opening up new areas of deeper and more remote waters for the first time.

The presence of gas can only be demonstrated by drilling a hole into the gas-bearing rock, and even after a series of appraisal wells have been drilled, estimates of the quantity of gas found must remain speculative. Only time will tell how much gas can be produced from a field once it has been developed, and changes in technology or the economics of gas production can change the definition of how much gas it is possible or profitable to produce. By their very nature, estimates of gas reserves must be unreliable guides, but they have to be made. The authoritative figures for Britain's gas reserves are those produced by the Department of Energy, which show the total amount of gas remaining in known discoveries at the end of 1977 as 1,546 billion cubic metres.

It has to be stressed that this is only an estimate of reserves in known discoveries, not an attempt to predict how much gas might be found in the future. Obviously, as time goes by, new gas discoveries are made, and the estimate of reserves rises. On the other hand, gas is being taken out of the fields already in production, reducing the remaining reserves. It is an indication of the continuing success of exploration that more gas is discovered each year than is produced, so that the Department of Energy's estimates of reserves keep rising.

At the end of 1973, the estimate of total reserves was 1,174 billion m^3; in the course of 1974, 34.8 billion m^3 was produced, but at the end of the year the estimate of reserves was 1,256 billion m^3. Over the four years from the end of 1973 to the end of 1977, a total of 149.6 billion m^3 was produced, but despite the loss of all this gas from the fields, estimates of reserves rose from 1,174 to 1,546 billion m^3. The limited usefulness and reliability of estimates of reserves may be demonstrated by the fact that the estimate for the beginning of 1965 would have been zero. Nevertheless, production began in 1967, and ten years later, a total of 241 billion m³ had been produced from the British sector of the North Sea. In 1965, our ignorance of gas reserves was total; now, it is only partial.

The reserves estimate of 1,546 billion m³ of gas remaining in known discoveries can be divided into three categories: 744 billion m³ of proven reserves, which are virtually certain to be producible; 354 billion m³ of probable reserves, with a better than 50 per cent chance of being producible; and 488 billion m³ of possible reserves, with a less than 50 per cent chance of being producible. The key word in these definitions is evidently "producible" — what can be produced with the present technology and would be profitable to produce at current costs and current price levels. Both technology and economics can change fairly rapidly. Gas is now being produced from the northern North Sea — not so long ago, even if it had been known that there was gas there, it would not have seemed possible to build the platforms and pipelines needed to produce it. If energy prices rise in real terms, then obviously gas fields which are now too small or too difficult to be profitable to develop might become worthwhile investments. Not only does the available knowledge about gas reserves continue to expand as exploration goes on, but the definition of commercially viable discoveries is also liable to change.

Even the most restricted definition of Britain's gas reserves, the proven discoveries which are almost certain to be producible, would suffice to maintain gas supplies at their present level for some years. Discoveries already made, but not considered so definitely worth producing, double the amount of gas we know about. The question of how much gas has not been discovered is logically impossible to answer, of course, but the Department of Energy has produced a speculative figure for all Britain's gas reserves, known and unknown, of 2,270 billion m³. The actual amount could be higher or lower; only time can tell.

The demonstrable and inevitable uncertainty of estimates of reserves shows the futility of just dividing the figure for reserves by the figure for annual consumption, and declaring that gas will run out in so many years. Natural gas is a finite resource, and production from the fields we now know about will cease one day, after a long slow decline which will last into the next century. The staggering success of exploration in finding so much gas in the years since 1965 makes it very hard to believe that no more fields remain to be discovered off the shores of Britain. Even if that is
58

the case, Britain is not necessarily entirely dependent on its own natural gas. The Frigg field has shown that gas can be imported from Norway, the Algerian LNG that gas can be imported from much further away. As a later chapter will show, it is possible to manufacture substitute natural gas from oil or coal. The idea of an "energy crisis" or an imminent "energy gap" does not seem as credible as it did when such gloomy notions were fashionable after the sudden increase in oil prices in 1973/74. In any case, gas supply is only part of the picture; there is also the question of gas demand.

Matching supply and demand

In the days when each town had its own gas works, making gas locally for a fairly limited and predictable market, it was fairly easy to produce the right quantity of gas to supply the customers' demands. Over the long term, rising demand could be met by expanding the capacity of the works, or a fall by shutting some of the plant. The variation in demand in the course of each day could be dealt with by using the familiar gas holders to store gas at times of low demand, to be drawn on when demand rose — in the evening, for instance, when people returned from work and lit their gas fires and cookers.

The same problems of variations in demand over the long term, between winter and summer and from day to day are still major preoccupations of the gas industry. With gas coming from fields far away in the North Sea, rather more sophisticated means of ensuring that customers will always get the gas they want, when they want it, have had to be adopted.

The problem of the long term match between supply and demand has gone through two phases. When gas first arrived from the North Sea, it arrived in such large quantities that strenuous efforts had to be made to find markets for it, so that an increasing supply dictated a rapid expansion. Now that the natural gas supply system is established, it has become possible to reverse the previous policy, and allow the growth of demand in selected markets to determine the growth of gas supply. Indeed, the principles of energy conservation should eventually lead to a falling demand in some markets, so that the amount of gas supplied each year will reach a stable level, then decline.

The initial phase of rapid growth was produced by a situation of increasing supplies, as the first gas fields in the southern North Sea were developed. Gas supplies quadrupled in ten years, a growth much faster than could be absorbed by the traditional market for

59

gas, which was mainly in the home and in specialized industrial uses. The new supplies of natural gas had to be sold, and the investment in transmission pipelines had to be paid for, so the gas industry sold much of the new gas to industry. At a time when total gas sales were less than 300,000 TJ a year, a contract was signed with ICI to sell 950 TJ a year, for use as a chemical feedstock. The price for large supplies of natural gas was much lower than the price of town gas, so it became a very attractive fuel for many industrial applications, and even for use in power stations.

Power stations of the conventional type generate electricity by burning fuel to produce steam. This is by no means a specialized use of fuel — the lowest grades of coal and oil, which could be used nowhere else, are good enough for producing the large quantities of crude heat needed. Two-thirds of the heat produced in power station boilers is lost in the process of electricity generation, so it is an inherently wasteful way to use fuels like gas, which can be better employed elsewhere. It is now accepted that gas should be reserved as far as possible for the premium market — that is, for those applications which make best use of the special qualities of gas, such as cleanliness and ease of control. The non-premium markets, where cruder fuel would do just as well, should be left to other fuels, which would not be capable of replacing gas in its specialized applications.

While North Sea supplies were building up, and a lot of gas was perforce sold to non-premium users, the premium market was also growing. In particular, sales of gas central heating were booming. By the middle of the 1970s, the importance of energy conservation was also generally recognized. The growth in premium demand, particularly in the domestic and commercial markets, made it possible to decide that the next major expansion in gas supplies, following the arrival of Frigg gas, should go chiefly to meet growing premium demand. If it proved necessary to reduce production from some fields until demand was large enough to absorb it, that would be better than supplying too much gas, which would have to be sold to non-premium users. The gas industry would prefer a long life supplying premium customers, rather than a few years of massive sales, from gas fields which will not last for ever.

Some of the contracts between British Gas and the companies which produce gas from the offshore fields have been re-negotiated to permit further flexibility in supply. As more gas arrives from the north, so it may be necessary to restrict production in the south, at least in summer, to avoid an excess of gas. The new contracts give British Gas more freedom to reduce the amount of gas it takes, thus helping to ensure that supply follows demand, rather than vice versa. The contracts specify how much gas British Gas must pay for in the course of a year, but British Gas can choose to pay for the gas and not take it, leaving it in the field for a later date when it might be needed.

The contracts with the producer companies also provide for the variation between summer and winter demand. Deliveries reach a peak in winter, as cold weather encourages increased gas use for heating. During the summer, fields can be closed down to allow maintenance work to be carried out on platforms. During the winter, the contracts allow deliveries to rise to 167 per cent of the average daily rate throughout the year. In this way, supplies from the North Sea move up and down, broadly in line with the seasonal fluctuation in demand.

Another means of ensuring that the gas industry can cope with peak demand in very cold winter weather is provided by the contracts between British Gas and some large industrial customers. These customers pay slightly less than other industrial customers for their gas, in return for allowing British Gas the right to interrupt their supply for a specified number of days each year. When these customers' supplies are interrupted, they switch to alternative fuels, thus releasing gas which can be used elsewhere. The interruptible customers thus help to guarantee continuous gas supplies to firm customers, like domestic users, when they need gas most in cold weather.

The need for special measures to increase gas supply in winter may be demonstrated by the fact that, in the late 1970s, the average amount of gas supplied over the course of a whole year was about 125 million m^3 a day, while on a very cold day in winter demand could rise to nearly 200 million m^3. The difference between the average demand and peak demand can only be met in part by increasing supplies from offshore and by the use of interruptible contracts. Storing gas which becomes available at times of low demand, ready for use when demand rises, has been practised for many years in gas holders. This method is still very useful in meeting local fluctuations in demand, storing gas in towns near to the customers so that it is possible to respond rapidly to changing circumstances. Other methods of storing larger quantities of gas are now also used, some to deal with the variation between winter and summer and some to meet the variation within each 24 hours — often, the highest demand early in the evening is three times greater than the lowest point, about 3 a.m.

The familiar gas holders which dominate the skyline of many towns, as they have done for more than a century, seem very large objects, capable of storing vast quantities of gas. While still useful in smoothing out daily fluctuations in demand, they are by no means large enough to provide storage for seasonal demand. Liquefied natural gas (LNG) has the great advantage that it occupies 600 times less space than the same quantity of gas in its gaseous state, so LNG tanks are widely used for bulk storage of natural gas — about 20,000 tonnes in each tank, equivalent to 25 million m^3 of gas. Tankers from Algeria deliver LNG to the storage tanks at Canvey Island all through the year. At a number of other sites around Britain, there are LNG tanks which are filled by liquefying gas from the transmission system in summer, when demand is low. When necessary, the LNG is re-gasified and fed back into the pipelines to provide an additional supply.

Another means of storing large quantities of gas is the use of underground salt cavities. Rock salt is impermeable to gas, so cavities in salt strata deep beneath the surface can be used as very large gas holders. Some countries, like Germany, have plenty of suitable sites for salt cavity storage, where the geological formations are appropriate, but there are few such likely sites in Britain. The only salt cavity storage site in Britain is near Hornsea, in Yorkshire, close to the coast. A number of cavities are being leached out of the salt stratum, by pumping sea water into the ground at the level of the salt layer. The water dissolves the salt, and brine is pumped out and back into the sea. Slowly, a cavity is formed, which can be used for storing gas for use at times of high demand.

The volume occupied by the quantities of gas which it would be useful to store, at normal pressures, can be reduced by liquefaction or by compression. High pressure storage vessels, generally spherical or cylindrical tanks, have been used to store gas near to centres of demand for many years. The transmission of gas in high pressure pipelines, operating at up to 69 bar, opened up further possibilities of storing gas at high pressure. Arrays of pipes were built, simply for holding gas which could be released into the supply system as required. Similarly, the transmission pipelines themselves hold a lot of gas all the time. By increasing pipeline pressure above the level needed for gas supply, more gas is packed into the pipelines, which can also be drawn upon as necessary.

These forms of storage, low pressure conventional gas holders, LNG tanks, salt cavities, high pressure vessels, pipe arrays and line packing, all play a part in maintaining the balance between supply coming from the North Sea and demand liable to wide fluctuations. This balancing act obviously depends on constant monitoring of all the sources of gas, the amount available in storage and the pattern of demand. At the national level, the terminals, transmission pipelines and offtakes are linked to the British Gas central control, which monitors the flow of gas throughout Britain and decides on the best strategy for supplying gas to the twelve regions. The regions each have a control room to operate their gas distribution networks from a central point. Information about the state of any part of the gas supply system is instantly available on visual display screens, linked to computers, making it possible for a few people in each control room to maintain the flow of gas to the customers, in the quantities they need, whenever they need it.

When the supply of gas from the North Sea was building up, and the process of converting Britain's gas supplies from town gas to natural gas was still under way, many gas works were modified to make substitute natural gas from oil. Substitute natural gas (SNG) was used to supplement North Sea supplies until the North Sea fields were producing enough to meet demand. When North Sea production begins to decline, whenever that might be, then SNG will probably again prove useful. At first, SNG is likely to be

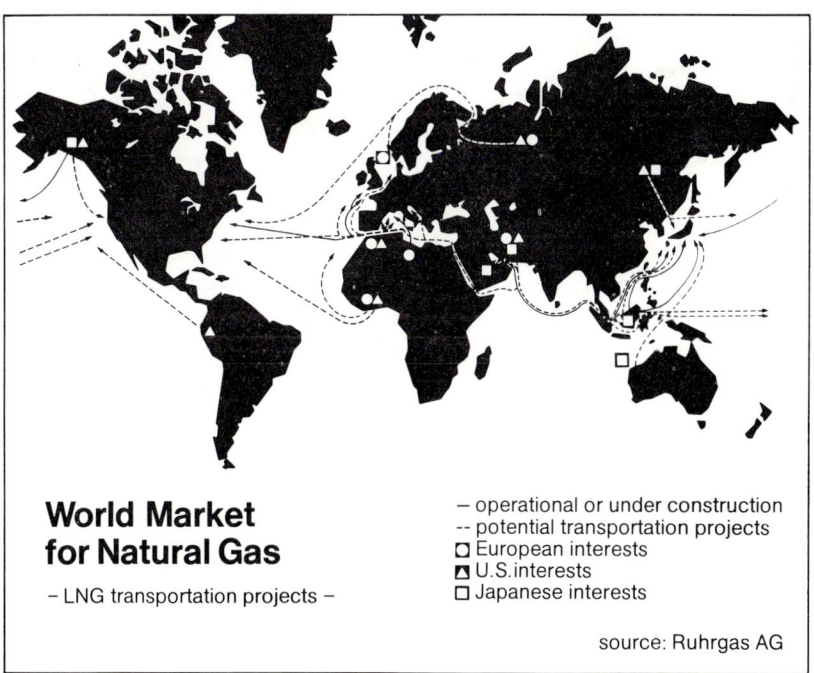

**World Market
for Natural Gas**

– LNG transportation projects –

— operational or under construction
-- potential transportation projects
□ European interests
◩ U.S. interests
□ Japanese interests

source: Ruhrgas AG

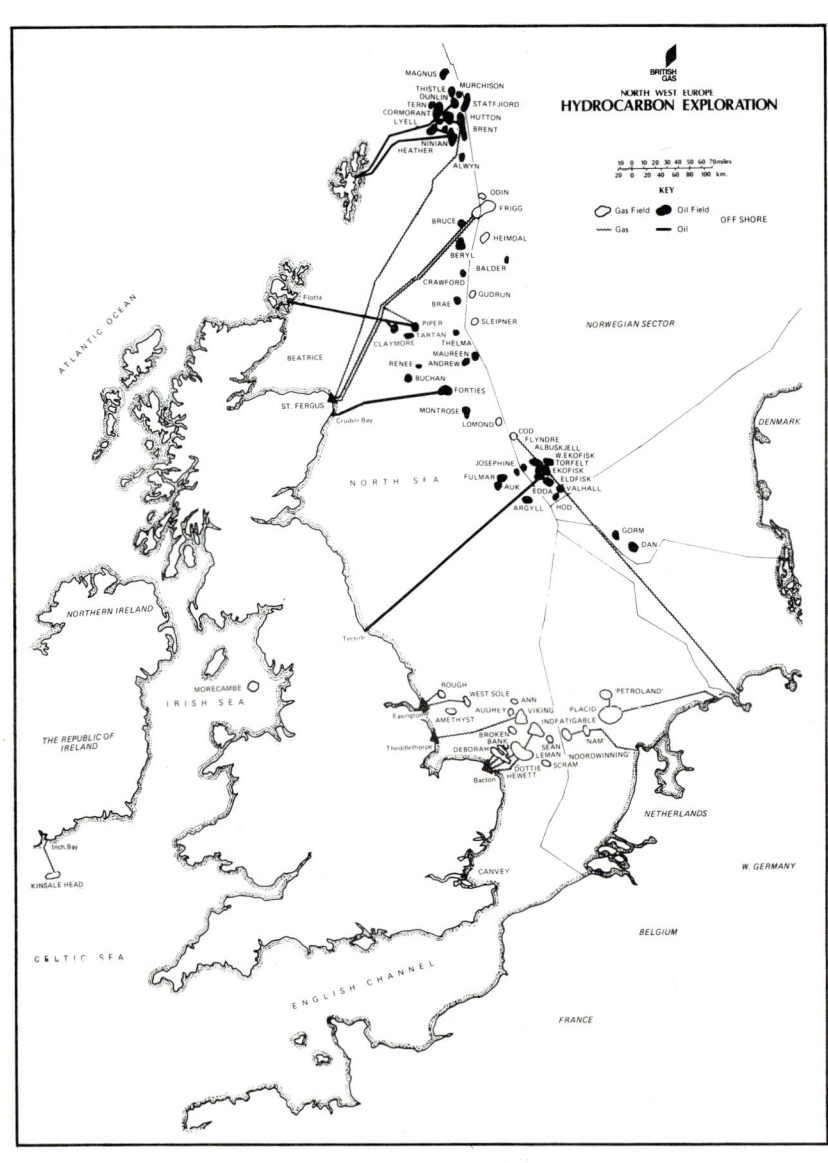

BRITISH GAS

NORTH WEST EUROPE
HYDROCARBON EXPLORATION

10 0 10 20 30 40 50 60 70miles
20 0 20 40 60 80 100 km.

KEY

⬭ Gas Field   ⬬ Oil Field   OFF SHORE
〜〜 Gas   ▬▬ Oil

MAGNUS
MURCHISON
THISTLE
DUNLIN   STATFJORD
TERN   HUTTON
CORMORANT BRENT
LYELL
NINIAN
HEATHER
ALWYN

ODIN
FRIGG
BRUCE
HEIMDAL
BERYL
BALDER
CRAWFORD
GUDRUN
BRAE
SLEIPNER
PIPER
CLAYMORE TARTAN
THELMA
MAUREEN
RENEE   ANDREW
BUCHAN
FORTIES
ST. FERGUS
Cruden Bay
MONTROSE
LOMOND
COD
FLYNDRE
ALBUSKJELL
W.EKOFISK
TORFELT
JOSEPHINE EKOFISK
ELDFISK
FULMAR   EDDA VALHALL
AUK
ARGYLL
HOD

NORWEGIAN SECTOR

DENMARK

GORM
DAN

Flotta
ATLANTIC OCEAN

BEATRICE

NORTH SEA

NORTHERN IRELAND

MORECAMBE
IRISH SEA

*THE REPUBLIC OF
IRELAND*

Teeside

WEST SOLE
ROUGH
'PETROLAND'
AUDREY ANN
AMETHYST VIKING PLACID
Easington   INDFATIGABLE
BROKEN
BANK   NAM
Theddlethorpe DEBORAH SEAN NOORDWINNING
LEMAN
DOTTIE SCRAM
Bacton HEWETT

NETHERLANDS

CANVEY

W. GERMANY

Inch.Bay
KINSALE HEAD

BELGIUM

CELTIC SEA

ENGLISH CHANNEL

FRANCE

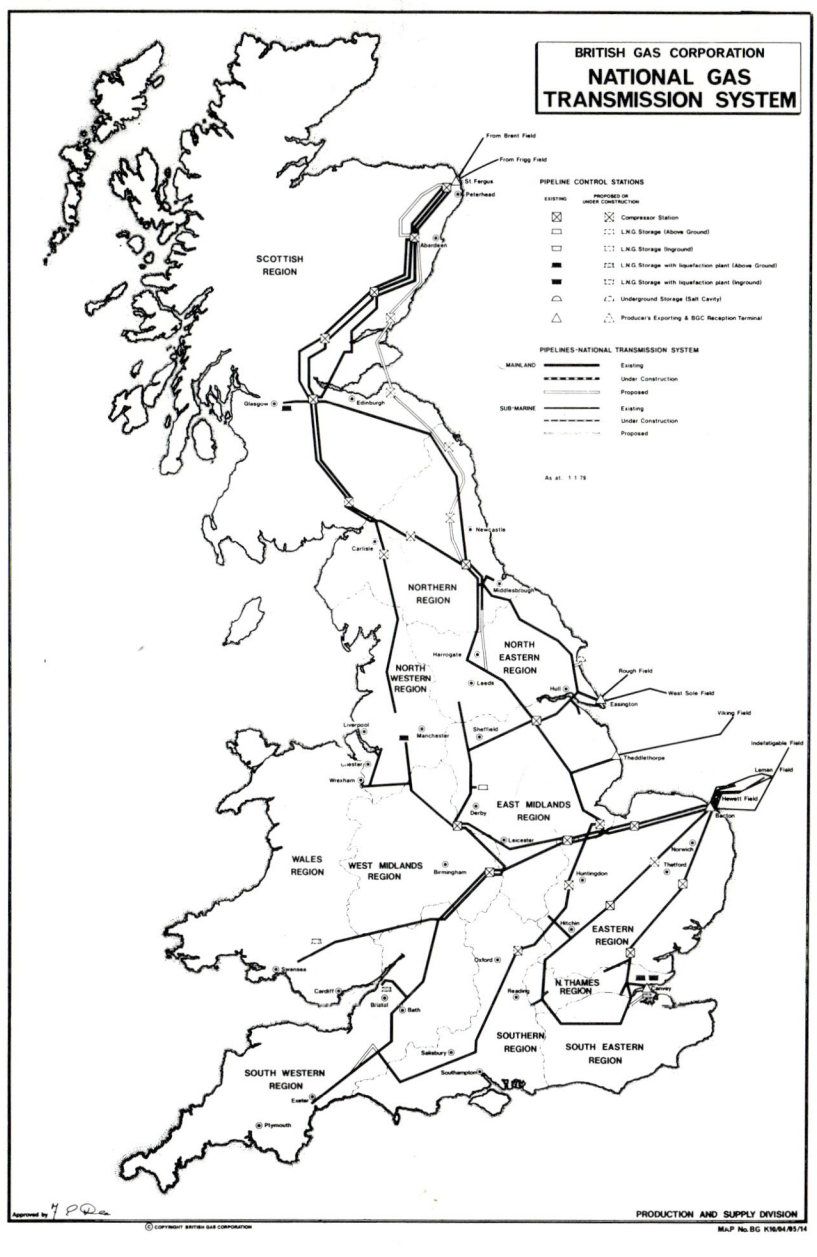

BRITISH GAS CORPORATION
NATIONAL GAS TRANSMISSION SYSTEM

PIPELINE CONTROL STATIONS

EXISTING	PROPOSED OR UNDER CONSTRUCTION	
⊠	⊠	Compressor Station
□	⊡	L.N.G. Storage (Above Ground)
□	⊡	L.N.G. Storage (Inground)
▬	⊡	L.N.G. Storage with liquefaction plant (Above Ground)
▬	⊡	L.N.G. Storage with liquefaction plant (Inground)
⌂	⌂	Underground Storage (Salt Cavity)
△	△	Producer's Exporting & BGC Reception Terminal

PIPELINES–NATIONAL TRANSMISSION SYSTEM

MAINLAND
————— Existing
--------- Under Construction
————— Proposed

SUB-MARINE
————— Existing
--------- Under Construction
————— Proposed

As at 1.1.79

PRODUCTION AND SUPPLY DIVISION

Map No.BG K16/04/85/14

Approved by

© COPYRIGHT BRITISH GAS CORPORATION

used just to top up North Sea supplies to meet peak demand in winter. Later, it could gradually take over from the North Sea as the principal source of gas for Britain. The manufacture of SNG could help to ensure that the gas industry will be able to continue to meet demand, particularly in the premium markets, far into the future. It could also mean that the gas industry will return to its traditional role as a manufacturing industry, making gas from coal as it did in the days of town gas.

5

MANUFACTURING GAS

Town gas from coal

The conversion of Britain's gas supply system and gas burning appliances from town gas to natural gas was completed in 1977. It is no longer relevant to go into too much detail about town gas, but to understand the way in which the gas industry has grown to its present shape and size, it is necessary to remember that for 150 years it was a manufacturing industry. Gas was made locally in gas works, rather than being a fuel made by nature and transported over long distances.

For most of the period when town gas was in use, it was made from coal, by a number of processes which differed slightly from one another, but which were all based on a simple principle: coal is a bulky, inconvenient fuel, which produces heat when it burns in a manner which cannot be precisely controlled; much of the heat content of coal can be harnessed in the form of a gas, a controllable fuel which can be piped to the user and just turned on and off. Town gas was originally intended to provide a fuel for lighting, an application for which coal was obviously unsuited, and gas was later adopted as a fuel for cooking and heating, as customers realized the advantages of replacing raw coal with the more refined fuel.

Town gas was made from coal by carbonization — when coal is heated in the absence of air, the volatile materials in the coal are driven off as a gas, leaving behind a residue of coke. This process was originally carried out in horizontal retorts, long low chambers with a flat base and a curved roof, perhaps 6 m long, 600 mm wide and 400 mm high, holding about 750 kg of coal. In time, retorts became larger, and were built vertically rather than horizontally, but the carbonization process stayed the same. Coal was fed in at one end of the retort, the opening closed and the retort heated.

8. *Beckton, once the largest gas works in Europe, employed 4,500 men to make town gas from coal. It supplied most of east and central London, in the days before oil gasification and natural gas.*

After 8 to 12 hours, carbonization was completed and coke was taken out at the other end of the retort. The crude coal gas produced during carbonization was piped away from the retort for purification.

The heating of the retort, to temperatures of up to 1,400°C, was accomplished by the use of producer gas. After each load of coal had been carbonized in the retort, the hot coke which was left was placed in a producer, underneath the retort. A stream of air was blown through the coke. The oxygen in the air reacted with the hot coke to form carbon monoxide, a flammable gas. Alternatively, air saturated with steam was blown through the hot coke, producing a mixture of carbon monoxide and hydrogen. Whichever process was used, the flammable gas from the producer was burned in combustion chambers to heat the retorts containing the next load of coal for carbonization. A simple and economical cycle was thus established, with coal giving coal gas and coke, and

68

coke giving producer gas to heat more coal.

The crude coal gas from the retort was a thick yellow vapour, containing many impurities such as tar, sulphur compounds and nitrogen compounds. These impurities were removed from the gas, many of them becoming the source of valuable by-products like ammonia, leaving a mixture of combustible gases ready for use as town gas. The composition of town gas varied according to the type of coal and the carbonization process used, but was typically a mixture of combustible gases like hydrogen, carbon monoxide and various hydrocarbons, principally methane. It could be diluted by mixing it with water gas — the gas made by blowing steam over hot coke — and it could contain a proportion of non-combustible gases, mainly carbon dioxide, oxygen and nitrogen. This mixture could be controlled, to produce a gas of fairly consistent quality, which would conform to the characteristics required for successful combustion in town gas burners. There were local variations in gas quality, and thus in calorific value, but the commonest calorific value was about 19 MJ/m^3 — we have already seen that the calorific value of the typical natural gas now in use is about 38.86 MJ/m^3. While methane is the principal component of natural gas, the proportions by volume of the various components of a typical town gas would have been:

	per cent
hydrogen	51.0
carbon monoxide	14.6
methane	19.1
other hydrocarbons	5.2
carbon dioxide	3.6
nitrogen	6.1
oxygen	0.4

This town gas had a specific gravity of 0.47, relative to air, which is hardly surprising as its most important constituent was hydrogen. The fairly large proportion of carbon monoxide in town gas made from coal had the unfortunate effect of making it toxic, and explains the popularity town gas once had as a means of suicide. Nearly 1,200 people died in 1963 as a result of accidental poisoning by town gas. Natural gas has many advantages over town gas, not least of which is the fact that it contains no carbon monoxide and is completely non-toxic.

The manufacture of town gas from coal was the basis of the operations of the gas industry from its origins early in the nineteenth century until the 1950s. It was displaced, not, as is often assumed, by natural gas, but by the manufacture of town gas from oil, which brought about a revolution in the gas industry which tends to be overlooked because it was so soon overshadowed by the discovery of gas in the North Sea.

Town gas from oil

The carbonization of coal to make town gas was a very efficient and successful process in its day, but it depended on the availability of suitable coal, at prices low enough to make gas manufacture competitive with other fuels. By the 1950s, suitable coal was becoming scarce and expensive, which stimulated research into other means of gas manufacture. One possibility was the development of better processes for making gas from coal, but a more immediately attractive prospect was the use of oil as a feedstock. Oil was then cheap and abundant, and the massive Middle East oil fields seemed to offer almost unlimited supplies costing less than coal for producing town gas. The gas industry embarked on a research programme which was followed very quickly by the widespread adoption of oil-based processes. The transformation of the industry was so rapid that it had almost completely changed its manufacturing operations in the few years before North Sea gas was discovered. In 1960, 90 per cent of the gas made in Britain was produced from coal; in 1967, when the first North Sea gas arrived, only 35 per cent was made from coal and oil was well on the way to taking over.

Before the 1950s, the only significant use of oil in gas manufacture was the carburetted water gas process, a hybrid process involving both coal and oil. Ordinary water gas, produced as part of the carbonization cycle by the reaction of steam with hot coke, is a mixture of carbon monoxide and hydrogen, with a low calorific value — what is known as a lean gas. If oil is sprayed onto a hot surface as part of the same production cycle, the oil vaporizes and cracks to give a mixture of gaseous hydrocarbons like methane, with a high calorific value — a rich gas. The resulting mixture of lean and rich gas, called carburetted water gas, has the appropriate calorific value for use as town gas of about 19 MJ/m^3. The cracking of oil in the carburetted water gas process to produce a mixture of gaseous hydrocarbons, and the mixture of rich and lean gases, are not dissimilar to the processes adopted in the 1950s when the gas industry began to change to processes using only oil. There are many such processes, but the principle generally followed in Britain was to produce a lean gas from oil, and then

70

9. Oil gasification was carried out at this catalytic rich gas plant in Hitchin. It was clearly a lot cleaner and more modern than coal carbonization at works like Beckton.

enrich it to give a mixture with the characteristics of town gas.

The production of a lean gas was generally achieved by the continuous catalytic reforming process, in which naphtha, a light oil, was reacted with steam over a catalyst at a temperature of 700-900°C and at high pressures, about 20-30 bar. The hydrocarbons in naphtha reacted with steam to produce carbon monoxide and hydrogen; carbon monoxide and hydrogen produced methane and water; carbon monoxide and water produced carbon dioxide and hydrogen. All three reactions took place in the reformer, the latter two being reversible reactions, and the composition of the final product could be controlled by altering the ratio of steam to naphtha, the temperature and pressure, and by the use of further processes to remove carbon dioxide or convert carbon monoxide. The gas produced by reforming could thus vary quite widely, but was typically a mixture dominated by hydrogen. It was a lean gas, with a calorific value of perhaps 11 MJ/m^3.

There were several ways of enriching this gas so that the town gas calorific value of 19 MJ/m^3 could be reached. The most straightforward was to add a gas which was much richer already, either liquefied petroleum gas (LPG) or natural gas. LPG is familiar as the fuel for cigarette lighters and in the gas cylinders used for camping stoves, consisting principally of propane or butane. Propane and butane are hydrocarbons which are gaseous at normal temperature and pressure, but easily liquefied under pressure, with calorific values of 94 and 121 MJ/m^3 respectively. Natural gas, with its calorific value of about 38 or 39 MJ/m^3, was also used for enrichment during the period when it was being produced from the North Sea, but town gas still had to be produced until conversion was completed. LNG from Algeria was originally imported for use in enrichment of town gas.

Natural gas and LPG are, as it were, ready-made rich gases. It is also possible to manufacture rich gases from oil feedstocks, and two processes were developed by the Midlands Research Station of British Gas to do this. The catalytic rich gas process is similar to continuous reforming in that steam reacts at high pressure with a light oil fraction like naphtha, in the presence of a catalyst. It is carried out at a lower temperature, about 500-550°C, and produces a gas mixture much richer in methane. The gas mixture produced also contains non-combustible carbon dioxide, which can be removed to give a higher calorific value. When used for enrichment, catalytic rich gas was produced with a calorific value of some 27 or 31 MJ/m^3, containing 64.4 or 74.0 per cent methane. Catalytic rich gas could also be subjected to a second process of re-

72

forming, increasing the proportion of hydrogen and carbon dioxide at the expense of methane, to give a town gas calorific value of 19 MJ/m^3.

The other rich gas production process was gas recycle hydrogenation, using heavier oil feedstocks instead of naphtha. The reaction of oil with hydrogen at high pressures and at a temperature of about 750°C yields a rich gas with a calorific value of 28 MJ/m^3. Conveniently, the hydrogen used in the reaction could be provided by the lean gas produced by continuous reforming. The rich gas produced by the gas recycle hydrogenator contained 41 per cent hydrogen and 31 per cent methane.

Whichever process, or combination of processes, was used to produce town gas from oil, the gas had the advantage over coal-based town gas in that it contained much less carbon monoxide, and was thus less toxic. The rapid adoption of these processes by the gas industry in Britain had the effect of reducing the number of gas poisonings from 1,193 in 1963 to 536 in 1967, when the first North Sea gas arrived. As natural gas is non-toxic, the number of accidental poisonings attributed to gas then fell even further, to 43 in 1975. So far as the gas industry was concerned, oil gasification had the advantages that it was more economical than coal carbonization, and did not produce the burdensome by-products, like coke, which were inseparable from carbonization. The catalytic rich gas process, in particular, was a great success not only in Britain, but also in Japan and Brazil, and more than 50 such plants were built to produce town gas from naphtha. The conversion of Britain's gas supply to natural gas has meant that there is now no need for town gas manufacturing processes, but the technology developed by British Gas has found another application, in the production of substitute natural gas.

Substitute natural gas

Natural gas is much richer than town gas, but not too much richer than the gas produced from oil by the catalytic rich gas and gas recycle hydrogenator processes. When Britain's North Sea supplies were building up to a level sufficient to meet peak demand, it was necessary to supplement them in some areas with a gas which would be interchangeable with North Sea gas. This was done by modifying some oil gasification plants to produce a richer gas, known as substitute natural gas or SNG.

Britain is fortunate in that it now has sufficient natural gas from the North Sea to meet demand for many years to come. Other countries are not so lucky. Parts of the USA suffer natural gas

73

10. Westfield pilot plant, in Fife, where substitute natural gas is made from coal as part of the research British Gas is carrying out to ensure supplies of gas far into the future.

shortages, particularly in winter, while Japan is almost entirely dependent on imported fuels. Such countries offer a market for SNG technology, and British Gas has profited from their problems, by making available the technology which was developed in Britain but is no longer needed here. In the long term, Britain will inevitably be in the same position as the USA, with declining indigenous supplies of natural gas, and may well start to use SNG on a large scale again. For Britain, this prospect lies many years in the future, but in the USA it is here already. British Gas continues to develop its SNG processes, so that it gains financially now, by selling its technology abroad, and at the same time builds up the expertise which it will need much later for SNG production at home.

The production of SNG from oil feedstocks, using the catalytic rich gas process followed by a further stage of processing to increase the proportion of methane in the gas, has been developed so successfully by British Gas that 15 plants are in operation in

74

Japan and the USA. In all, these plants have the capacity to produce 42 million m^3 a day of SNG — as much as the average daily flow from the Frigg field. Since 1973/74, all oil prices have risen considerably, but naphtha prices have risen more than most, as it is in great demand. The high prices paid for naphtha as a feedstock for chemical manufacture make it likely that it will become a very expensive basis for SNG production. British Gas has therefore extended its research into SNG processes to allow a wider range of oil products to be used. By using different catalysts, it is now possible to produce SNG from oil fractions heavier than naphtha, such as kerosine, but even these may become scarce one day.

It seems likely that coal will continue to be available when oil production declines. The implication for the long term future of the gas industry in Britain is that, when SNG is needed on a large scale, coal will be a more attractive feedstock than oil. It has already been mentioned that in the 1950s, when the limitations of coal carbonization were becoming apparent, the gas industry became interested in processes for making town gas from oil and in better ways of gasifying coal. Just as the oil-based town gas processes proved relevant to SNG manufacture, so the coal research was later to be applied to SNG, opening up the prospect of a long term future for gas, even when natural gas and oil are no longer available.

Electricity has been described as "coal by wire" — the coal used in power stations produces a more desirable form of energy. Similarly, the coal used in carbonization was transformed into a controllable fuel which could be delivered by pipe. On the other hand, carbonization also produced by-products like coke and tar and a good deal of pollution, so that only half the heat content of the coal was actually delivered as town gas. This is better than the 30 per cent efficiency of conventional power stations, but could be improved if coal could be gasified completely, leaving behind only ash. Carbonization also had the disadvantage of being limited to certain types of coal. A more efficient means of making gas from a wider range of coals was offered by the Lurgi process.

The Lurgi process was developed in Germany in the 1930s, and modified by research in Britain in the 1950s, as a means of producing town gas by total gasification. In a Lurgi gasifier, low-grade coal reacts with steam and oxygen at a pressure of 20-30 bar. It is a continuous process, with coal fed into the top of a generator, and ash removed at the bottom, giving a lean gas containing methane and a high proportion of carbon monoxide. After further treatment to convert carbon monoxide to carbon dioxide and to remove much of the carbon dioxide and sulphur compounds, the result is a lean gas which can be enriched and used as town gas. The advantages of the Lurgi process, using low-grade coal and producing no coke, led to the building of two Lurgi gas works in the 1960s, at Westfield in Fife and Coleshill in the West Midlands. The success of oil gasification processes, then Lurgi's rival for town gas manufacture, limited the adoption of Lurgi technology in Britain to these two works, and the introduction of natural gas led to their premature redundancy.

The Lurgi plant was still standing, however, and British Gas had built up an unrivalled store of experience and expertise in coal gasification. It became clear that the research and the plant could still be put to good use, if they were adapted to the investigation of making SNG from coal. The Lurgi plant at Westfield was no longer needed to make town gas, so it became available for research, concentrating on its operation at high temperatures, to economise in the use of steam, and on the development of the slagging gasifier, in which the coal ash left after gasification fuses and is removed as a molten slag. For British Gas, this research could have no immediate relevance, as long as the North Sea lasts, but there is a much more pressing need for such technology in the USA. There, a natural gas shortage seems imminent, while there are extensive coal reserves, which can be produced cheaply by strip mining. A consortium of American companies, led by Conoco, sponsored a series of research programmes at Westfield, which demonstrated that the slagging gasifier could cope with a wide range of coals. American coals, including some with a high sulphur content, were imported, and successfully gasified, and the SNG produced fed into the local gas distribution network.

Westfield has proved quite conclusively that SNG, interchangeable with North Sea natural gas, can be produced from coal. The slagging gasifier has been shown to deal with many types of coal quite unsuitable for carbonization, and interest in the technology has developed to such an extent in the USA that the Department of Energy there is considering plans to build a commercial SNG plant in Ohio. Meanwhile, the research at the Westfield pilot plant goes on, with the intention of extending the range of coals which the gasifier can accept still further. The next stage is a composite gasifier, which can use fine coals as well as lump coal. By the time that SNG is needed in Britain, it should be possible to produce it from nearly any type of coal at an efficiency of 70 per cent.

The demand for SNG in the USA has had the incidental advantage that American interests have been willing to provide the entire cost of the research into the slagging gasifier. Britain has thus acquired its SNG technology, with the prospect of royalties from abroad if the process is adopted by foreign gas companies, at their expense. For Britain, the production of SNG from coal is a long-term insurance policy, against the day when the North Sea is no longer able to meet our need for gas. That day may be a long way off, but the technology is available now, and it works. Other forms of SNG manufacture have been suggested, such as using nuclear power for the electrolysis of water to produce hydrogen, which could then be used to gasify coal, but they remain speculative. They may have become practical possibilities by the time when Britain needs SNG, but whatever technology is then used, it is clear that gas will be available as long as there is a demand for it.

6

THE GAS INDUSTRY AND ITS CUSTOMERS

British Gas

Since 1973, the supply of gas in Great Britain has been the responsibility of a single organization, the British Gas Corporation. The rights and duties of British Gas were laid down by Parliament in the Gas Act 1972, under which the old system of the Gas Council and twelve area gas boards was replaced by a unified gas industry. The main duty placed on the Corporation is to develop and maintain an efficient, co-ordinated and economical system of gas supply for Great Britain and to satisfy, so far as is economical, all reasonable demands for gas in this country.

The gas industry in Britain was nationalized in 1949, when all the private and municipal gas undertakings were taken into public ownership. The gas industry in Northern Ireland, the Channel Islands and the Isle of Man was not affected by nationalization, and remains in the hands of a number of separate undertakings. In Great Britain, on the other hand, the unification of the industry has made it the largest single integrated gas undertaking in the world. British Gas is the only gas undertaking to be involved on a large scale in all aspects of supplying gas, from exploration to looking after its customers' gas appliances; in most countries, gas production and transmission, local distribution, the sale of appliances and fitting and servicing are carried out by several different organizations. The integration of all these aspects in one Corporation has made British Gas, with assets of £2,000 million, one of the largest businesses in Britain.

A member of the Government, the Secretary of State for Energy, appoints the chairman, deputy chairman and members of the British Gas Corporation, but thereafter the Government's role is limited to giving occasional instructions on broad matters of national policy, such as setting target levels of profit for British

Gas. The running of the industry is left to the Corporation, which appoints the chairmen of the twelve regions of British Gas, and these regions take care of the day-to-day business of dealing with individual gas consumers and supplying them with gas. The headquarters of British Gas determines policies and administers all those matters which affect the industry as a whole, particularly the national gas transmission system.

Policy-making on the highest level is the concern of the Corporation, which has five full-time members and several part-time members. The part-time members are prominent trade unionists, businessmen and chairmen of gas regions, who bring expertise from many different fields to the running of British Gas. The full-time members are the heads of five of the divisions of British Gas headquarters — Finance, Marketing, Production and Supply, Resources and External Affairs, and Personnel. The functions of Finance and Personnel divisions are self-explanatory; Marketing is concerned with selling gas and with selling and servicing appliances; Resources and External Affairs involves all sources of gas, both British Gas's own exploration for gas and the purchase of gas from other producers, as well as international relations; Production and Supply is mainly an engineering division, controlling gas supply operations such as terminals, compressors and transmission pipelines. There are two other divisions, Research and Development, with four research stations, and Economic Planning, as well as various miscellaneous functions like the legal department and public relations gathered together as the Chairman's Office.

Headquarters is responsible for bringing natural gas across the country, but its distribution to the customers through the gas mains is the business of the regions. Headquarters may decide the price of gas, but the regions actually read the gas meters and send out the gas bills. Bills can be paid, and gas appliances bought, at some 970 gas showrooms, run by the regions. This devolution of affairs as they affect each customer is necessary in a business covering the whole country, with 14 million customers to deal with.

The supply of gas in Great Britain is the monopoly of British Gas. Even before nationalization, when local gas works run by private companies or the local authorities supplied town gas, the customer had no choice as to whom he bought his gas from. Even then, it was obviously absurd to have several companies each digging holes in the road to repair separate gas mains serving the same street, so local monopolies were established. Now that all our gas comes from the North Sea, the need for a unified gas supply

79

system is still more apparent. This argument does not necessarily apply to selling and servicing gas appliances, and here there is competition between British Gas and private installers. Before nationalization, many gas undertakings had showrooms and departments to carry out fitting and servicing, and these were taken over together with their gas supply operations.

British Gas is the largest retailer of gas cookers, fires and water heaters in Britain, but it sells only a minority of the gas central heating systems installed. The gas consumer is free to choose any supplier of gas appliances, at least in theory gaining the benefits of competition. The same could be said of the apparent monopoly of gas supply, where competition is provided, not by a rival gas company, but by coal, oil and electricity.

Other organizations
While British Gas, as the sole supplier of natural gas in Great Britain, dominates the gas industry, many other companies are involved, as suppliers and competitors in some aspects of the business. Most of the equipment used in gas transmission and distribution is bought by British Gas from a wide range of engineering companies, and much of the work of construction and pipelaying is contracted to other companies. The laying of a major transmission pipeline, for instance, involves the purchase of pipe from British Steel or from foreign manufacturers, and fittings such as valves, from many other companies. British Gas plans the route for the pipeline and supervises its construction, but the actual work of digging trenches and laying pipe is carried out by contractors. Such projects do not occur regularly, and the contractors have the pool of skilled labour and the machinery to do the work when it is required.

A relatively small group of companies manufactures most of the familiar brands of domestic gas appliances. As British Gas is the largest retailer of gas appliances, much of the manufacturers' output is sold to British Gas, and there are close links between the two sides in the development and testing of appliances. Not all appliances are sold by British Gas, however, and in the field of central heating particularly, their market is much wider. The trade organization representing manufacturers of appliances and of equipment used in the supply of gas is the Society of British Gas Industries.

The installation of gas appliances, as has been said, is not solely undertaken by British Gas. There are thousands of other installers, most of them operating on a fairly small scale within a limited area. On the fringes of the trade, there are some people who carry out installation, probably together with a bit of plumbing and a bit of building work, without being sufficiently qualified to do a good job. The importance of proper gas installation, for the customers' safety, led to the establishment of the Confederation for Registration of Gas Installers, known as CORGI. CORGI keeps a register of the qualified gas installers in all parts of the country, and periodically inspects the work of its members, to ensure that they are maintaining acceptable standards. Some gas installers who are not on the CORGI register may be perfectly competent, but others are Jacks-of-all-trades, who might not be relied upon to do as good a job as a specialist in gas fitting. The importance of making sure that gas installation is done only by specialists makes it advisable for customers to employ only CORGI members or British Gas for such work.

Safety is also the concern of the law. The Gas Safety Regulations require that all gas appliances should be installed by competent persons, so that do-it-yourself gas fitting is against the law. A breach of the regulations is punishable by a fine of up to £400, although, inevitably, most incompetent gas installations are only discovered when it is too late, and something has gone wrong. Like any other fuel, gas is safe when used properly, but unskilled work can, at the worst, cause explosions or the risk of suffocation due to inadequate flues. The Gas Standards Branch of the Department of Energy is the governmental body with a general responsibility for gas safety, and for prosecutions under the Gas Safety Regulations.

The Gas Standards Branch is also responsible for monitoring the quality of the gas supplied, by ensuring that British Gas complies with the legal requirements for gas quality, pressure, uniformity of calorific value and odorosity. In this way, it acts as an independent check, ensuring that consumers receive gas of as good a quality as they are paying for. This applies particularly to calorific values, which vary from one place to another, depending on the source of the natural gas supplied; the minimum calorific value of the gas must be declared, and adhered to, so that customers receive the heat content specified. Similarly, the accuracy of gas meters is the concern of Gas Standards Branch, which carries out independent checks of meters to determine if they are recording gas consumption to within specified limits of accuracy.

The interests of gas consumers, in the broadest sense, are looked after by the Gas Consumers' Councils, whose members are appointed by the Department of Prices and Consumer Protection.

The National Gas Consumers' Council puts forward the point of view of gas users on general matters of policy, such as gas prices, and is consulted by British Gas. Twelve regional Gas Consumer's Councils, corresponding to the regions of British Gas, take care of matters affecting individual consumers, including the investigation of complaints which consumers have not been able to resolve with gas regions. The Gas Consumers' Councils have also undertaken the task of monitoring the operation of the Code Of Practice on disconnection, which lays down the circumstances in which British Gas can cut off gas supplies to consumers who have not paid their bills.

The Institution of Gas Engineers is the professional association through which gas engineers exchange technical and scientific information. It plays a major role in laying down the recommended practices for gas engineering which are followed in Britain, and maintains links with gas engineers abroad, through the International Gas Union.

Working in British Gas
British Gas employs 100,000 people, which appears to be a large number, but is low relative to its 14 million customers and to the 140,000 employed in the gas industry in the early 1950s, when gas was still made from coal. The introduction of natural gas has reduced the number of people employed in gas manufacture and supply, and has also led to a change in the kind of people employed. There is no longer any need for the unskilled manual labour of shovelling coke, or even for much of the unskilled labour of digging trenches for gas mains. Mechanization and the increasing complexity of gas supply operations have led to a smaller work-force, with a higher level of specialised skills. Some of the old occupations involved in gas manufacture have disappeared, but the range of activities of the gas industry has expanded, so that it now employs more diverse skills in everything from exploration to appliance servicing. Manual workers are now outnumbered by white-collar staff, and the traditional occupations like distribution craftsmen and fitters have been joined by jobs once unknown in the gas industry, like geophysicists, computer programmers and even an archaeologist.

This expansion of the range of activities within British Gas has mainly affected a relatively small proportion of its employees, who provide the more esoteric special skills. The largest group of employees, as might be expected, is still to be found in customer service, where there are 30,000 fitting and repairing gas appliances. Another 20,000 work in transmission and distribution, laying and maintaining the system of pipelines, mains and services which supplies the gas. The next largest group is perhaps not so obvious — 12,000 in customer accounting. This part of the operation is equally important, as it is their job to collect the money from the customers. With 14 million customers, each with a meter that has to be read four times a year, who then have to be sent a bill, often followed by a reminder before payment is received, customer accounting needs a large staff. The fourth major group of employees are the 11,000 in marketing and sales, many of whom work in gas showrooms or behind the scenes, selling gas and appliances. These four main groups make up the majority of employees and cover the major activities of British Gas. The remainder are engaged in a multitude of occupations, many concerned with the administration of an industry of this size, such as the personnel staff looking after recruitment and training and industrial relations. Planning, finance, research and all kinds of engineering each need their specialist workers to enable the industry to function.

The shift in technology in the gas industry has been closely reflected in the number of employees. In 1968, there were 69,000 manual workers and 53,000 staff. By 1978, the number of manual workers had fallen to 40,000, while the number of staff had risen slightly to 59,000. Over the same period, the number of customers and the amount of gas sold had both risen considerably, so that productivity had shown a marked improvement, which still continues. Productivity can usefully be measured in terms of the number of employees required to provide service for a given number of customers, or of the amount of gas sold relative to the number of employees: in 1968, there were 9.3 employees for every thousand customers, in 1978 only 6.9; the amount of gas sold per employee rose from 3.6 TJ in 1968 to 16.1 TJ in 1978.

These rough and ready indicators of productivity demonstrate the effect of the revolution brought about by natural gas, and a great improvement in efficiency in many areas. The ending of gas manufacture, the completion of conversion to natural gas and the introduction of computers caused many jobs to disappear, many of them unskilled manual jobs or tedious clerical occupations. This trend had started before North Sea gas, during the earlier transition from coal to oil for making town gas, but with the establishment of natural gas a greater degree of stability has been achieved. The rapid fall in the number of employees has been halted, as British Gas has settled down to a level of about 100,000 employees. The extension of gas supplies to new customers and the replacement of old mains may create more jobs in distribution in the future, while the increasing number of gas appliances sold, and their greater complexity, could lead to a growing need for fitters to install and service them.

There is thus the possibility of a slight expansion of the gas industry, but even to remain at the same level needs a continuous supply of new recruits. Apprentices, with CSE or SCE qualifications in scientific or technical subjects, with an aptitude for practical work, are recruited to learn the crafts of gas distribution and servicing. Dealing with customers' enquiries, accounts and appliance sales create opportunities for a variety of jobs in offices and showrooms. The development of gas technology and of office technology means that in all areas of the gas industry there is a trend away from unskilled work and repetitive clerical tasks towards the employment of a diversity of specialized skills.

Gas consumers

The rapid rise of gas sales since the arrival of North Sea gas has been the result of an increase in the number of gas consumers, an increase in the amount of gas each customer uses and an expansion of the market for gas into new areas. The change over the decade following the year when gas from the North Sea was first supplied was particularly marked:

	1967/68		1977/78	
	Number of customers (thousands)	Gas sold (PJ)	Number of customers (thousands)	Gas sold (PJ)
Domestic	12,560	280	13,963	735
Industrial	78	96	70	682
Commercial	572	67	483	184
Total	13,210	443	14,516	1,601

The domestic market naturally accounts for the vast majority of customers, and is now taking just under half of the total amount of gas sold. There is still some potential for expanding gas sales to the domestic sector. Of the 19 million households in Great Britain, nearly 14 million have a gas supply. About 2½ million households are in areas too remote to be connected to the gas distribution system in the foreseeable future or are in some other way debarred from using gas, but this leaves another 2½ million homes where gas could be provided, if required. On past evidence, increasing the

84

number of domestic customers is a slow business: there were already 12 million as long ago as 1957. Increasing the amount of gas used by each existing customer is a more likely form of expansion.

In 1967/68, average gas consumption was 22.3 GJ for each domestic customer; in 1977/78, the average was 52.6 GJ. In other words, that mythical creature, the average customer, now uses twice as much gas as was the case in the days of town gas. Much of this increase can be explained quite simply. The price of gas, relative to other fuels, has fallen; central heating has become the norm rather than the exception; gas is the most popular fuel for central heating. Traditionally, gas was used for cooking, water heating and gas fires, but the principal form of heat in most homes was the coal fire. Coal has been displaced by gas as the principal domestic fuel, and gas central heating consumes much more gas than the traditional uses. A typical family's cooking might use 8.4 GJ a year, and a gas fire in the living room 26.4 GJ a year. Central heating, depending on the size of the house and the climate, can consume between 80 and 125 GJ a year.

Gas central heating was almost unknown in the 1950s, but by the late 1970s more than 5 million homes had gas central heating, and more than half a million more systems were being sold every year. About 11 million gas cookers were in use, and 13 million space heaters, like gas fires and unit heaters. Sales of nearly 600,000 cookers a year, and over 900,000 space heaters a year, were to some extent accounted for by the replacement of old appliances, so the total number in use did not rise as rapidly as was the case with central heating. The growth of central heating, where some 70 per cent of all systems sold were fuelled by gas, was the major factor in the rise of gas to be the most important fuel in the domestic market. By the end of the 1970s, gas provided more than 45 per cent of all the heat supplied to the domestic market.

Over these ten years which saw the rapid expansion of gas sales to domestic customers, the number of commercial gas consumers fell slightly, while sales to the commercial sector nearly trebled. The most important part of the commercial market for gas used to be in hotels and catering, where gas was used, much as it is in the home, for cooking, heating and hot water. Hotels, restaurants, pubs and clubs used 32.4 PJ in 1977/78, representing 18 per cent of the total commercial gas market, but had by then been overtaken by educational users. Schools, colleges and universities also produced a striking rise in gas sales, to 34.3 PJ, as gas displaced coal and oil as the preferred fuel for heating. A similar development occurred in hospitals, which used 24.4 PJ in 1977/78, five times as much as they used ten years before. Offices and shops have also become major gas consumers. Some individual commercial customers consume very large quantities of gas: the London Hilton Hotel uses 105 TJ a year, about as much as 1,000 houses with gas central heating; all the heating and catering at Gatwick airport is gas-fired, consuming 190 TJ a year.

The industrial gas market is much more markedly a matter of relatively few customers each using very large quantities of gas. Over the decade in question, there was a slight fall in the number of industrial customers, while sales multiplied seven-fold. In the days of town gas, there was a fairly limited industrial market for gas, as it tended to be more expensive than coal or oil except for specialized applications where a clean controllable fuel was particularly needed, or the direct heat of a gas flame had to be applied. The most important sectors of the industrial gas market were in engineering and shipbuilding and in the iron and steel industry.

The introduction of natural gas changed all that. Not only was the price low — the average price for industrial users fell from 62.9p per GJ in 1967/68 to 28.1p per GJ in 1972/73 — it was also available in vast quantities. Natural gas could be used as a feedstock in the chemical industry, and it could compete with coal and oil as a source of heat on a large scale. Gas was sold by contracts negotiated individually with each customer, many of them on an interruptible basis. Such sales to non-premium customers like power stations helped to absorb the sudden influx of gas and to give the gas industry the flexibility it needed to be able to meet peak demand.

This phase of unrestricted growth ended when the build-up of supplies from the fields in the southern North Sea flattened out. The gas industry spent two or three years waiting for Frigg gas to arrive in 1977, before expansion could begin again. Prospective industrial customers had to be asked to wait, as the premium market took up available additional supplies. The price of gas in industrial contracts was linked to the price of oil, its major competitor, and as oil prices rose, so did the price of gas. By the end of 1978, a company signing a contract for a gas supply could expect to pay an average of £1.43 per GJ.

This rise in prices helped to prevent the profligate use of gas in non-premium applications. British Gas adopted a policy of concentrating its gas sales on the premium market, where the special characteristics of gas would be fully exploited, leaving the provision of crude heat to crude fuels. The decision to expand supply

86

only so far as is justified by demand, rather than by selling the maximum amount of gas which can be made available, has helped to ensure that the gas reserves we have can be consumed in the optimum way for as long as possible. At the same time, British Gas has played an important part in encouraging energy conservation in industry, establishing the School of Fuel Management, which provides education in practical methods of saving energy, for factory managers and other representatives of industry. The Technical Consultancy Service of British Gas was set up to give expert advice to industrial and commercial customers on making the best use of gas, and has helped many of them to achieve impressive fuel savings.

Virtually every sector of British industry uses gas to some extent, either as a means of heating its premises or in manufacturing processes, but the largest gas consumer uses gas, not as a source of heat, but as a feedstock for transformation into other products. In the chemical industry, natural gas is used as a feedstock for making ammonia, which is a major component of fertilizers. About a third of industrial gas sales go to the chemical industry, which used only 4 PJ in 1967/68, but used 233 PJ in 1977/78. The next most important customer, the iron and steel industry, was far behind, using 60 PJ in 1977/78, followed by food, drink and tobacco producers, and then by vehicle manufacture.

The price of gas

Large industrial and commercial gas customers, who use more than 10.5 TJ a year, negotiate individual contracts with British Gas, at prices related to the market levels of competing fuels, particularly oil. The price of gas to the domestic consumer, on the other hand, is fixed by published tariffs, revised regularly. There are two types of tariff, the pre-payment tariff for customers who pay for gas as they use it with a coin-in-the-slot meter, and the credit tariff for customers who pay for gas after they have used it, when they are sent a bill every three months.

Tariffs are based on the principle that they should reflect the cost of supplying gas. In effect, the cost of gas supply falls into two parts. On the one hand, there are certain costs which would be incurred for every customer, irrespective of the amount of gas consumed. The repair and replacement of gas mains, service pipes and meters, the reading of meters and the preparation and postage of bills would have to be carried out even if the customer used hardly any gas, just as they would be if a lot of gas was consumed. On the other hand, there are the costs related to the actual amount

87

of gas consumed. For this reason, tariffs fall into two parts. A fixed standing charge or a higher price for the initial quantity of gas consumed is intended to recover, at least in part, the fixed costs, with a lower rate charged for subsequent quantities. Tariffs for pre-payment meters tend to be higher than for credit meters, simply because pre-payment meters are more expensive to install, maintain and collect money from.

Gas tariffs also differ in various parts of Britain. When gas was made locally, there were differences in the costs of producing it, related to the availability of coal and concentration of customers; where customers were scattered over a wide area, there were higher costs for distribution through longer mains. The introduction of natural gas from a common source has helped to diminish these regional differences, and a multitude of tariffs has been reduced to just three zones. Customers in Scotland, Wales, southern England and East Anglia pay the same tariffs as each other, while tariffs are a little lower in the north of England and lower still in the Midlands.

In Britain, gas meters measure the quantity of gas delivered in terms of cubic feet, while customers pay for gas according to the amount of heat supplied, measured in therms. These units will be replaced when the metrication of gas finally takes place. The volume of gas will presumably be measured in cubic metres, as is already the case for the government's statistics for North Sea production. The therm will also have to be replaced. The SI equivalents are the joule and its multiples, but it has been suggested that Britain might invent its own new unit, equivalent to 100 MJ. As one therm is equal to 105.5 MJ, the new unit would be of roughly the same size as the traditional unit. It is not yet certain if or when this change will take place.

Whatever unit gas is measured in, it is difficult to talk of an average price, due to the regional variations and the different types of tariff. Some idea of the movement of gas prices might be gained by considering changes in the average revenue received by British Gas for each gigajoule of gas sold to domestic customers. In 1967/68, this average revenue was 97.5p per GJ. Most of the gas sold in that year was town gas, but the progress of conversion meant that customers were moving to natural gas at prices lower than those for town gas at that time. Despite inflation, there had been little change in average revenue by 1973/74, when it was £1.04/GJ. Prices had in fact been kept down to an artificially low level as part of the then government's policy of price restraint. When this restraint was lifted, there was a rapid rise in gas tariffs, as there was in the price of most other commodities. In 1977/78, the average revenue from much larger domestic gas sales was £1.75/GJ.

Looked at in isolation, this increase seems steep, but it should be borne in mind that it happened at a time of rapid inflation, when wages and prices were both rising fast. The Retail Price Index, which compares the prices of all the goods which a typical family might buy, showed that overall prices had more than doubled in the five years from January 1974 to January 1979. Over the same five years, the index for gas rose by 76 per cent, compared with rises of 150 per cent for coal and 165 per cent for electricity. Wages and old age pensions, too, went up faster than the price of gas, so that the cost of gas in real terms, that is, relative to the amount of money customers had available to spend, actually declined. Of course, arguments about prices in real terms make little impression on customers, who are more concerned with the sums they have to pay out when the gas bill arrives. Nevertheless, the general acceptance of the fact that gas is a less costly fuel for home heating than electricity, oil or solid fuel, has helped to make gas the most widely-used source of heat in the home.

Where the money goes

As we have seen, gas tariffs reflect the cost of supplying gas, in that they fall into two or more parts related to fixed costs and running costs. This relationship between prices and costs also applies on the larger scale, as British Gas is supposed, under the Gas Act, to operate in such a way that it covers its costs and can make an allocation of its financial surplus to building up reserves. The government can also issue instructions to British Gas on the level of profit it is supposed to make, as in 1979/80, when it was set a target of a 6½ per cent return on turnover. The price of gas has to be fixed at a level high enough for British Gas to cover its costs, and still make a surplus, the return, out of its total income, or turnover.

An example from a recent year, 1977/78, will show where all the money which gas customers paid to British Gas went. The total revenue from gas sales was £2,199 million, most of which, £1,287 million, came from the 14 million domestic customers. British Gas also received revenue from the sale of oil from the North Sea oil fields in which it has an interest, from sales of gas appliances, from fitting and servicing appliances and from such sources as overseas consultancy work and royalties paid for the use of British Gas technology. This brought the total turnover to £2,568 million.

The sales of gas appliances just about broke even, and service work made a small loss, so all the profit for that year could be attributed to sales of gas itself, which is clearly the most important part of the operation. The trading profit is the amount left from the total turnover, after paying all the operating costs of running the gas supply system. In 1977/78, these operating costs came to £2,254 million. The first cost which British Gas had to meet was that of the natural gas itself, bought at the shore terminals from the offshore producer companies for £460 million. In other words, less than a fifth of the operating costs of the gas industry was represented by gas; the rest was made up of the costs associated with bringing the gas from the shore terminals to customers throughout the country.

Some of the most significant of these other costs which the gas industry incurred in supplying gas are fairly obvious. The materials and services used in gas supply, like pipes and the rest of transmission and distribution hardware, cost £168 million. Contractors were paid £170 million for their work in building, pipelaying and other work which is contracted out. The salaries and wages of 100,000 employees, together with associated costs like the employer's contribution to insurance and pensions, came to £501 million. In 1977/78, £239 million was taken by the costs of conversion to natural gas and the ending of the usefulness of the old gas works, but this is a cost which will not arise again, as conversion ended in 1977. The last major operating cost was £341 million for depreciation — the money set aside for renewing and replacing existing assets, such as gas mains, which all wear out in time. The cost of replacing such assets is rising continuously, but it is essential that they should be replaced when they are no longer adequate, so a large sum has to be allowed for depreciation.

The trading profit of £314 million was left from the £2,568 million turnover after paying all these operating costs involved in supplying gas. But there were still more costs which had to be provided for out of this trading profit, and ultimately out of the money paid by gas consumers. Interest payments amounted to £133 million. These interest payments had to be made as a result of the large sums which British Gas had borrowed in previous years. The bulk of the national gas transmission system was built in just a few years, at great expense, and an investment of this size could only be financed by borrowing. The loans are now being paid off, but meanwhile there is still a large bill for interest payments. After interest, there was a pre-tax profit of £180.3 million, from which £76.4 million was allocated for taxes, leaving just £103.9 million as the profit after tax, which could be transferred to the reserves.

In 1978, the reserves of British Gas amounted to £313 million. This is the money which British Gas will need to make new investments, to expand and develop the gas supply system. If the gas industry is to take advantage of new technology, and new sources of gas like the Morecambe field are to be exploited, then this will cost money, which is what the reserves are for. Profits are necessary to build up reserves, and in 1977/78 profits, after interest and tax, represented just 4 per cent of the total amount paid to British Gas by its customers. The income of British Gas is spent on supplying gas, on paying for past investments in the gas supply system, on reserves for future investments, and on making sure that the system is maintained in good condition, so as to be safe and efficient.

The safe use of gas
Gas is a fuel. It would be of little use if it was not combustible, but the heat it supplies to keep us warm and well-fed can also be a source of danger if gas is wrongly used. Fire, the proverb says, is a good servant but a bad master, and much the same applies to gas, and to all other fuels. Gas is very safe when used properly, and it has a very good safety record, as good as any other fuel, but this should not blind us to the need to take due precautions.

Conversion to natural gas has effectively ended what used to be the greatest risk from town gas, poisoning by unburnt gas. When town gas was made from coal and contained a large proportion of carbon monoxide, there were literally thousands of deaths by gas poisoning each year, either accidental or deliberate, as so many people used town gas as a means of suicide. This is no longer possible with non-toxic natural gas. In addition, conversion made a further great contribution to safety, as every gas appliance in Britain had to be inspected to be sure that it could be converted to burn natural gas, and this check helped to reduce the number of appliances in poor condition because they were old and had never been serviced.

Once, a gas leak was poisonous. Now, there is only the risk that gas and air might mix in the right proportions to make a flammable mixture, and find a source of ignition — a spark, a lighted match or any other flame. The rapid burning of a gas/air mixture in a confined space can cause a pressure rise sufficient to break windows or cause structural damage — an explosion. Considering the number of people who use gas, the number of gas explosions is quite low. In 1977/78 there were 99 explosions, and three deaths from explosions. This figure was below the average for the previous five years, 108 explosions and ten fatalities a year.

It is interesting to note that when the explosions over the five years 1972 to 1977 are analysed, mains and services were the source of only 32 per cent of the incidents, and more than two-thirds resulted from a source within customers' premises, the meter, installation piping or appliances. A considerable proportion, 12 per cent, were the result of accidental interference with the gas supply, such as excavations in the street, 14 per cent resulted from faulty workmanship, 13 per cent from customers using their appliances wrongly, and 17 per cent from deliberate malpractice — vandalism, attempts to steal from meters, trying to use gas without paying for it. In other words, many gas explosions could have been avoided with a little care and common sense, and by ensuring that gas installation and fitting is only carried out by qualified gas fitters.

Gas explosions have to begin with gas leaks mixing with air, so they can often be prevented if people take note of any smell of gas. The distinctive and unpleasant smell is put into natural gas so that leaks will not be ignored. If you smell gas anywhere, at home or in the street, report it immediately to the local office of British Gas — the emergency telephone number is in the telephone directory, under "Gas". If the leak is at home, turn off the gas supply at the meter. It ought to go without saying that you should not smoke, light matches or switch electrical appliances, including lights, on or off — this could cause a spark. British Gas makes no charge for investigating gas leaks and making them safe.

The other potential hazard from gas, poisoning by the products of combustion, can also be avoided by a little care and by leaving gas fitting to professionals. If gas appliances are correctly installed, used properly and serviced regularly, gas will burn safely and the products of combustion will be no danger to health. Problems arise when flues are blocked and the products of combustion cannot escape. Annual inspection of gas fires, central heating and water heaters ensures that appliances are in good working order, with clear flues. The provisional figure of 80 gas poisonings in 1977/78 is higher than it need have been if only all gas appliances had been properly installed and maintained.

This account of dangers from gas should serve to point out the importance of using it carefully. The same applies to all other

92

fuels, and it might be relevant to point out that there are several thousand accidental deaths in the home every year. Relative to the total number of accidental deaths, and to the 14 million homes using gas, the figures for accidents involving gas are very small. It has been said that the chance of death in a gas explosion is about the same as the chance of being struck by lightning. Small as the risk is, it can be reduced still further, by reporting gas leaks, using gas with care and leaving gas fitting to competent specialists.

7

INTERNATIONAL GAS

World gas reserves and production

The difficulties of estimating Britain's gas reserves, and the limitations on the available estimates, have already been explained. It can easily be imagined how much more difficult it is to produce estimates for the whole world's reserves. Many areas have been little explored, and many countries produce no figures for their resources. Only the most rough and ready guess can be made about the reserves which are known about, and the spread of exploration to more and more remote areas rapidly makes any estimate obsolete. Nevertheless, some attempts have been made.

The World Energy Conference in 1976 produced estimates of reserves of the three major fossil fuels. World recoverable reserves of solid fuel were put at 16.4 million PJ; world proven recoverable reserves of crude oil at 3.9 million PJ; and world proven recoverable reserves of natural gas at 2.3 million PJ, or 63,171 billion m^3. Whatever the deficiencies of these estimates, such as the fact that many countries did not provide information, they at least give some idea of the relative size of coal, oil and gas reserves, with coal being much the largest.

A more recent estimate of proven gas reserves at the end of 1977 is shown in the table. The total of 62,630 billion m^3 is much the same as the World Energy Conference figure. One point which is immediately apparent is that the USSR has a third of the world's natural gas reserves, twice as much as the next most important country in the list, Iran. Not so apparent is the fact that one of the most important consumers of energy, Japan, is absent from the list. It is also clear that much of the world's proven reserves is in countries whose climates make it unlikely that there will be much demand for gas heating, and where there is little industrial demand for energy. While these estimates have been left

World Proven Natural Gas Reserves, End 1977

	billion m³	
North America	8,400	
USA		5,900
Canada		1,700
Mexico		790
South America	1,920	
Venezuela		1,190
Argentina		240
Bolivia		180
Asia (excluding USSR) & Oceania	3,340	
Australia		800
China		600
Indonesia		640
Pakistan		420
Malaysia, Brunei		470
Middle East	16,900	
Iran		10,600
Saudi Arabia		1,860
Qatar		1,660
Kuwait		1,050
Iraq		770
Abu Dhabi		520
Africa	5,900	
Algeria		3,470
Nigeria		1,460
Libya		800
Western Europe	3,700	
Netherlands		1,700
UK		800
Norway		650
West Germany		180
Italy		190
Eastern Europe	22,470	
USSR		22,000
Poland		130
Romania		120
WORLD TOTAL	62,630	

behind by exploration, which has increased the proven reserves of countries like Mexico, they at least show that natural gas is found in nearly every part of the world, but not necessarily distributed in accordance with demand.

The developed countries, where industry and urban life have created the demand for large quantities of energy, tended to base their industrialization on the availability of coal. In much of the world, natural gas has only been exploited on a large scale relatively recently, being previously unknown or ignored as an annoying substance produced together with oil, fit only for flaring. There is thus an obvious discrepancy between natural gas reserves and potential demand. The figures for natural gas production in 1977 also show an uneven distribution. World production of natural gas was 1,400 billion m^3, and the leading producers were:

	billion m^3
1. USA	564
2. USSR	346
3. Netherlands	97
4. Canada	74
5. UK	40
6. Romania	36
7. Iran	22
8. Mexico	21
9. West Germany	19
10. Italy	14
11. Venezuela	12
12. Algeria	10
13. France	8
14. Poland	7
15. Hungary	6

On a world scale, proven reserves are sufficient to support production at the present level for 45 years, and as in Britain, exploration is likely to add to the proven reserves. On a national scale, comparison of reserves and production figures shows that some countries are using their gas reserves much faster than others. In countries like the USA, Romania and France, proven reserves are equivalent to ten years' production or less. On the other hand, reserves in Iran, Mexico, Venezuela and Algeria could last for much longer, and are far larger than could be absorbed by local demand.

Some 70 per cent of the world's energy is consumed in Europe

and North America, while Japan, virtually without indigenous resources, is another major consumer. For a long time, these three
areas of high energy demand have been supplied with oil brought
by tanker from the countries of the Middle East and South
America, which are rich in oil but have little local demand for it.
When oil was relatively cheap, there was little interest in doing
the same with natural gas. Furthermore, the technology which
would make it possible to transport gas over long distances in the
same way as oil was not available at prices low enough to make gas
competitive. The sudden rise in oil prices in 1973/74 changed all
that.

For years, the gas found together with oil in the Middle East
had been considered a nuisance, to be flared in order to get rid of
it. Giant gas flames have burned in the deserts of Arabia for year
after year, as part of the process of oil production. When oil was
cheap, there was little point in doing anything else. The recent rise
in oil prices changed all this and brought about a demand for gas
as a commodity in international trade. Flaring still goes on, but
the oil-producing countries of the Middle East are now planning to
sell their gas rather than waste it. They have not been converted to
the importance of energy conservation, but they can see that
making use of gas is now technically possible and potentially
profitable.

The prospect of profit from natural gas has not only increased
interest in gas which was only too easy to produce, and had to be
flared, it has also stimulated exploration and development in areas
which were once considered too unpromising or too inhospitable.
At the same time, the rapid advance of technology has opened up
offshore areas for exploration. In the early 1950s, the only offshore gasfields producing were in the calm, shallow waters of the
Gulf of Mexico. The discovery of gas in the North Sea began a
move into deeper and rougher waters, and into much more remote
areas. Gas has now been found in places as diverse as the Canadian
Arctic and the Gulf of Thailand.

Gas from the North Sea is fairly close to its potential market,
but gas from these more remote locations has to be transported
over long distances to a market. There is, after all, little point in
producing gas unless it can be delivered to a customer who is
willing to pay for it. Carrying gas over these long distances can be
done in two ways, by pipeline or by liquefied natural gas carrier.

Long distance pipelines
Large diameter high pressure pipelines have long been used to

carry natural gas within countries like the USA and Canada, from the gas fields to the cities which provide most of the demand for gas. In Britain, the 5,000 km of pipelines which make up the national transmission system bring gas to every part of the country. Britain and, until recently, the USA have been unusually fortunate among the industrialized countries in having enough natural gas from their own resources to meet home demand. Other countries in continental Western Europe have had to become dependent upon imports of gas from countries which can produce more gas than they need, through a network of pipelines.

France, West Germany and Italy all produce some natural gas, but far too little to meet the demand for gas in those countries. Countries like Belgium, Finland, Spain and Switzerland are entirely dependent on imported gas. There are four main sources for gas imports into the European gas grid, LNG from Algeria and Libya, the Dutch onshore fields, the Norwegian sector of the North Sea, and the USSR. Between them, these sources provided 72 per cent of the gas used in continental Western Europe in 1977, the most important being the gas fields in the Netherlands, providing 54 per cent of the total. The Dutch fields, particularly the Groningen gas field, are so prolific that the Netherlands has been able to sell gas to other countries, despite using very large quantities of gas at home. Norway has plenty of gas in the North Sea, but little domestic demand; in any case, it would be very difficult to construct a pipeline to the Norwegian coast. Norway has thus sold gas from the Frigg field to Britain, and from the Cod and Ekofisk fields to a group of European countries. A pipeline has been built to a shore terminal near Emden in West Germany.

Natural gas from the USSR travels much greater distances. The Soyuz gas pipeline, 1400 mm in diameter, runs 2,700 km from the Orenburg gas fields in the Urals into Eastern Europe. Gas from deep inside Russia supplies Soviet allies like East Germany and Czechoslovakia, and is also sold to Western Europe, through links with the European gas grid. Parts of this grid are still under construction, but it will eventually link West Germany, Belgium, France, Switzerland, Italy and Austria to gas brought by pipeline from the Netherlands, the USSR and the Norwegian North Sea, and gas landed by LNG carrier from North Africa. The USSR also supplies gas to Finland.

Two major additions to the European gas grid have been proposed, both very ambitious and perhaps speculative. Before the fall of the Shah, it had been agreed that gas from Iran would be sent to Austria, West Germany and France, by way of 6,000 km of
98

pipeline, running over the Caucasus, through the USSR and Czechoslovakia into the West. The revolution in Iran halted the supply of gas from Iran to the USSR, which had already been established, and has cast considerable doubt on plans for Iran to supply 5 per cent of Western Europe's gas by 1985.

The other proposal is for two pipeline links across the Mediterranean. Algeria has been forced by its geography to export its gas as LNG, but direct pipeline links with Europe would enable it to sell more gas. One route would run from the gas fields in the Sahara, across Tunisia to the shores of the Mediterranean. From Tunisia, six to eight undersea pipelines would run to Sicily, and more pipelines from Sicily to the Italian mainland, where gas would be fed into a large diameter pipeline running 1,000 km to the north of Italy, linking with the existing grid. The other route would cross the Mediterranean from Algeria to Spain, at water depths of more than 2,000 m, and through Spain to France.

These projects are subject to considerable uncertainty, in one case political, in the other technical, and even if they do go ahead, it could be many years before any gas is delivered. Nevertheless, they show the international scale of the trade in natural gas, and the way in which pipelines can link many countries, and bring gas thousuands of kilometres from the fields to the customers. Elsewhere in the world, similar long distance pipelines have been proposed: a 1,500 km pipeline from Australia's North West Shelf offshore fields to Perth; another running 1,350 km from Mexico's Chiapas field to the Texas border, which might supplement the USA's declining indigenous supplies; the Alaska Highway Gas Pipeline, from Alaska through Canada to the gas customers of the Western USA; and perhaps the most ambitious, a 3,750 km pipeline from Melville Island in the Arctic Ocean to Longlac in Ontario, at an estimated cost of $7,000 million.

Pipeline projects on this scale are likely to play an increasing role in maintaining gas supplies to Europe and the USA, but they may not be enough. Gas demand in Western Europe is forecast to increase from 170 billion m^3 in 1977 to 250 billion m^3 in 1985; demand in the USA is also rising, while the long-established fields in states like Texas and Louisiana are barely sufficient to provide adequate supplies, particularly in winter; Japan is clearly not likely to be linked by pipeline to any other country. Short sea crossings by pipeline are feasible, as in the North Sea, but over longer distances LNG carriers provide the answer.

Liquefied natural gas in international trade

Pipelines are inflexible, tying gas producers and purchasers to one another. They are also expensive, and impracticable over long distances involving the crossing of oceans or of several countries, all of which have to agree to their construction. As natural gas can be liquefied, and thus reduced in volume sufficiently to become a suitable cargo for carriage by sea, LNG carriers are playing a growing part in the international gas trade. As with most other aspects of energy, the oil price rise of 1973/74 was the crucial point at which the economics of LNG began to seem attractive to many potential purchasers and producers.

The first LNG project, to transport gas in barges up the Mississippi from the Gulf Coast of the USA to Chicago, was put forward in 1952, but there was little serious interest in LNG transport until the late 1950s. The Gas Council, the predecessor of British Gas, together with American companies, converted a small

11. *An LNG carrier unloads its cargo of liquefied natural gas at a terminal. LNG is stored at a temperature of –161°C, in the tanks connected by insulated pipes to the jetty.*

100

freighter to carry LNG. In 1959, the "Methane Pioneer" crossed the Atlantic, making the world's first LNG ocean crossing. Experimental voyages with the "Methane Pioneer" demonstrated that LNG transport was a practical possibility. What was needed in the producer country was a pipeline to carry gas from the fields to the coast, where there would be a terminal for liquefying the gas and loading it onto carriers. LNG carriers would have to be specially built, with tanks that could hold a liquid at $-161°C$, surrounded by insulation to keep it cold, using any gas that boiled off as fuel. At the other end, the country receiving the gas would have a terminal where carriers could unload and LNG could be stored and regasified for transmission by pipeline.

The first commercial LNG scheme started operating in 1964, linking Algeria and Britain. Algeria had found gas at Hassi R'Mel in the Sahara, and Britain needed natural gas, as the North Sea fields had not yet been discovered. Terminals were built at Arzew on the Mediterranean coast and at Canvey Island in the Thames estuary, and two LNG carriers, the "Methane Princess" and "Methane Progress", were constructed. These have been shuttling back and forth between Algeria and Britain since 1964, with a carrier arriving at Canvey every five days on average, without incident. The contract between Britain and Algeria expires in 1980, and may or may not be renewed.

Algeria was thus the first LNG exporter, and also found customers in France, Spain and the USA. Libya supplies LNG to Spain and Italy. The growing demand for gas in Europe, and the construction of the European gas grid, have made it likely that imports of LNG from North Africa will increase in the 1980s. New LNG terminals are planned for Zeebrugge in Belgium, Wilhelmshaven in West Germany, Emshaven in the Netherlands and Montoir in France, in addition to the existing terminals at Le Havre and Fos in France, La Spezia in Italy and near Barcelona in Spain. This development is predicted to lead to Algeria supplying 18.5 per cent and Libya 1.3 per cent of the total natural gas demand in Continental Western Europe by 1985. The construction of these terminals would also allow Europe the option of purchasing LNG from more distant sources, when it becomes available.

Europe is able to draw on gas from many sources, including LNG. Japan is not so fortunate, and its isolation and lack of indigenous energy resources have made Japan the world's largest consumer of LNG. Much of the LNG is used for electricity generation, a use which is justified by the Japanese on the grounds that it causes much less of an environmental problem than using nuclear power or coal. LNG is imported by a number of separate gas and electricity companies, and there is no unified national gas transmission system. By the late 1970s, Japan was importing LNG from Alaska, Brunei, Abu Dhabi and Indonesia. Further supplies were expected from Sarawak, Qatar and, before the revolution, Iran, to produce a situation by 1985 in which LNG imports would mostly be used to provide 15 per cent of Japan's electricity.

In the longer term, Japan hopes to import LNG from Australia and Siberia. The Australian North West Shelf gas fields are large enough to supply Western Australia and to allow exports on a large scale. The USSR has confirmed gas reserves of 825 billion m^3 in Yakutia, in Siberia (larger than all Britain's proven North Sea reserves). A project to link the gas fields to a terminal near Magadan, on the Sea of Okhotsk, whence LNG could be shipped out, is being studied by the USSR and Japanese and U.S. energy companies. Projects on this scale may be some years in the future, but as Japan is already supplied with LNG brought 6,000 km from Alaska and 4,000 km from Indonesia, they may well come to fruition, so pressing is Japan's need for energy.

There are now LNG carriers with a capacity of 125,000 m^3, which could supply in one year enough gas for heating and cooking in 800,000 homes. The transport of such large quantities of gas as LNG makes it possible to link the most remote gas fields with the major centres of gas demand, in Europe, Japan and North America. The price of all forms of energy seems likely to continue to rise, making LNG increasingly attractive to both producers and purchasers. In time, LNG could become a commodity traded and used as widely as oil, as the gas producing countries seek markets for gas which was once either ignored or flared, and the industrialized countries seek new sources of the energy on which their prosperity depends.

8

A SHORT HISTORY OF THE GAS INDUSTRY

The Chinese are said to have used natural gas as long ago as the third century A.D. In the seventeenth and eighteenth centuries, experiments with gases, including gas made from coal and natural gas, were carried out by scientists throughout Europe. But for all practical purposes, the history of the gas industry is generally considered to have begun in 1792, when William Murdoch used the gas he produced from coal, to light his home in Redruth, Cornwall.

Murdoch was an engineer, working for the Birmingham engineering company of Boulton and Watt. He persuaded them to exploit the use of coal gas for lighting factories, thus bringing gas out of the realm of experiment and into practical application. The next stage in the development of gas was its application for street lighting, a much-needed advance in the days when even the main streets of great cities were poorly and fitfully illuminated by oil lamps. A German who adopted the name of Frederick Winsor had seen experiments in producing gas from wood in France, and he arrived in London in 1803 with a revolutionary idea. He suggested that gas could be manufactured at a central point, and distributed through mains under the street, like water, rather than being confined to large factories and houses, each needing their own gas-making plant.

Winsor was more a talented entrepreneur than a scientist or engineer, but his abilities as a publicist and campaigner provided the driving force for the expansion of gas manufacture from private use to become a public service. He began with lectures and demonstrations in theatres, and in 1807 provided the first public exhibition of gas lighting for the street, by piping gas from his own house to lamps in Pall Mall, London. After overcoming considerable opposition from those with interests in lamp-oil and from those who feared the dangers of this novel and untried fuel, Winsor succeeded in winning parliamentary approval for his plans. In 1812, the Gas Light and Coke Company was granted a charter to supply gas in London and Westminster.

The technical problems of gas manufacture and distribution were rapidly solved by more practical men than Winsor, who established the principles of making gas from coal in horizontal retorts, purifying it, storing it in gas holders and supplying it through mains. By 1819, London had 460 km of mains and 51,000 gas lights in the streets. The advantages of well-lit streets were so obvious that gas companies were set up in most large towns, and by 1849 there were nearly 700 gas undertakings in Britain. The spread of gas was so rapid, and so little controlled, that often several companies competed with one another to supply gas to the same area. Walworth Road, in South London, for example, had four sets of mains run by separate companies on each side of the road! Eventually, Parliament acted to introduce some order into the situation, in London in 1860 and elsewhere in 1871, by giving gas undertakings local monopolies for particular districts, and introducing independent checks on the quality of the gas they supplied. The supply of gas has thus been a monopoly, so far as the individual customer is concerned, for more than a hundred years. Originally, gas supply was in the hands of private companies, but many towns followed the example of Manchester, which in 1843 set up the first municipal gas undertaking, controlled by the local authority.

In the USA, the use of gas for street lighting began in 1817 in Baltimore, and the idea spread as rapidly as it had in Britain. Also in the USA, the first short-lived experiment in using natural gas commercially followed the drilling of a shallow well in Fredonia, New York, in 1820. Gas lighting began in 1819 in Paris and Brussels, and by 1850 town gas had been introduced into most European countries. By the 1850s, gas was also being made in Canada, Brazil, Mexico, Australia and India. In the 1860s, the exploitation of natural gas really began in the USA, following the discovery of oil and gas fields in Pennsylvania. Pipelines were laid to transmit natural gas from the fields to the cities, where it was used mainly for street lighting. In Britain, natural gas was found at Heathfield, Sussex, in 1869, but the amount of gas produced was only enough to supply the lamps at the local railway station.

During the nineteenth century, many types of gas cooker, fire and water heaters were invented, but the most important use of gas remained the lighting of streets and homes. At first, light was provided by a simple flame at the point where gas emerged from a pipe through a burner nozzle. In the 1850s, Robert Bunsen developed the burner named after him, which allowed air to be mixed with gas in controllable proportions before it reached the point at which it was to be ignited. One of Bunsen's pupils, Carl Auer, was working in his laboratory at Heidelberg in Germany when he noticed that a sheet of asbestos impregnated with certain chemicals would produce an intense glow when placed in the heat of a gas flame. In 1887, Auer, later given the title Carl Auer von Welsbach, introduced the incandescent gas mantle, which produced a bright light by directing the heat of a flame onto a fabric impregnated with chemicals to make it glow. The gas mantle not only displaced the simple gas flame as a means of lighting, but also allowed the gas industry to compete with the newly introduced electric light.

Eventually gas gave way to electricity for lighting streets and homes, but the competition between the two fuels was long and intense. As late as 1933, half the street lights in London were gas and half were electric; a few gas lights still survive here and there, lighting old streets and consciously picturesque Victorian pubs. The gas industry found that as soon as homes were connected to an electricity supply, electric light replaced gas. Worried that it might lose all its customers to the new fuel, which might displace gas altogether, the gas industry tried strenuously to resist its encroachments. In the 1930s, electricity often gained entry to a house when the family decided to buy a radio. To hold back this process, which might be just the first step before people went on to buy electric lights, fires and cookers, the gas radio was produced. It was not a success, as you may have noticed from its conspicuous absence today!

The future of the gas industry did not lie in lighting, or in attempts to prove that electricity was unnecessary by selling bizarre appliances like gas radios and gas irons and gas hairdryers. Its survival was assured by selling gas for the purposes to which it was best suited, heating, cooking and water heating. Fairly efficient appliances for these tasks were available in the 1880s, and their widespread use was greatly helped by a rather obscure invention, the pre-payment gas meter. The pre-payment meter, which delivered a measured quantity of gas when a penny was placed in the slot, was introduced in 1887. It made it possible for gas to reach even the poorest families, who could pay for it a penny at a time. While the pre-payment meter hardly seems a particularly staggering innovation, it nevertheless opened up the mass market of ordinary working households to the gas industry.

In the period after the introduction of electricity, the gas industry managed a remarkable expansion: in 1882, there were 500 authorised gas undertakings, selling 1.8 billion m^3 of town gas; in 1912, 826 undertakings sold 5.5 billion m^3. The First World War brought a further expansion, with the by-products of gas works in demand for making explosives. In the period between the wars, this growth continued, with production reaching 8.9 billion m^3 in 1939, but to some extent it disguised the problems of the gas industry.

Gas was still produced in the 1930s by the traditional methods of coal carbonization in horizontal or vertical retorts, by hundreds of separate gas undertakings, some private and some run by local authorities. The more progressive companies, like the Gas Light and Coke Company which served most of north London, invested in new plant and concentrated their gas manufacture in a few large works, thus ensuring considerable economies: Beckton works was the biggest gas works in the world, occupying over 200 ha, employing 4,500 people and supplying east and central London. The smaller companies, on the other hand, had a few hundred customers served by an old and inefficient works and ten or twenty kilometres of mains, and were unable to invest in better plants, or to operate more efficiently by making gas on a larger scale.

It was plain that there would have to be some closer integration of gas undertakings to bring the whole industry up to date and to allow the smaller works to be phased out. A number of schemes for reorganization were put forward during the 1930s and the Second World War, but when the Labour government was elected in 1945, committed to public ownership of the most important sectors of the British economy, nationalization of the gas industry was seen to be inevitable. The Gas Act 1948 brought the whole industry into public ownership. Some 1,100 gas undertakings, operating 1,050 gas works, were grouped into twelve area gas boards, with the Gas Council as a central co-ordinating body.

The new organization, which came into existence in 1949, took over the problem of modernizing and rationalizing the chaotic structure of the gas industry. Each of the old undertakings had fixed its own tariffs, so that the Eastern Gas Board alone found over 400 tariffs in force in its area. Separate gas distribution systems had to be linked, so that manufacture could be concentrated on the most efficient works and gas could be carried by local pipeline grids over a wide area. Massive investment was necessary to make up for the neglect of the industry during the Depression of the 1930s and the damage caused during the war. Despite these

106

efforts, the gas industry's future was in doubt.

Gas manufacture depended on coal supplies, and coal suitable for carbonization was becoming scarce and expensive. Meanwhile, the electricity industry, which had also been nationalized, was building new power stations to burn cheaper grades of coal and oil, linked by a national electricity grid. Between 1950 and 1957, the price of gas rose by 51 per cent, while electricity prices only went up 17 per cent. Many observers thought that gas was finished, a relic of the nineteenth century like steam trains. Electricity seemed to have the image of the all-purpose fuel of the future, particularly as the 1950s were the time when nuclear power was regarded most optimistically as a source of unlimited cheap energy. Gas appliances too were beginning to look archaic beside the attractive modern designs of electrical alternatives.

As it happened, the introduction of nuclear power stations was not to be a quick and simple matter, and it is still bogged down in controversy. More to the point, the people who worked in the gas industry were not convinced that gas was obsolete. To compete with electricity in the home, appliances had to be designed to appeal to a newly-affluent public, offering greater convenience and looks that would fit into modern homes. The research and development facilities of the Gas Council backed the gas appliance manufacturers in the introduction of a succession of new ideas. Automatic ignition for cookers, eye-level grills, convector fires, thermostatically-controlled water heaters and oven timers were all launched in the 1950s, followed in 1961 by wooden-cased gas fires. The impact they made, particularly the new gas fires designed to look like attractive modern furniture, rather than functional metal boxes, is hard to imagine, now that they are all so commonplace. At the time, they contributed to a revival in appliance sales, and thus in the use of gas.

More fundamental changes were necessary in the actual process of gas manufacture if the industry was to escape a decline into obsolescence. The economies possible by shutting down small inefficient gas works and integrating gas distribution networks were near their limit by 1960. The 1,050 works of 1949 had been reduced to 428, but these fewer, larger works were still dependent on expensive grades of coal. A complete revolution in the processes of gas manufacture was needed if gas was to be produced more efficiently, at a competitive price.

Three possibilities were under investigation by Gas Council scientists and engineers in the 1950s: more efficient ways of gasifying coal, the gasification of oil and the use of natural gas. Research into coal gasification led to the adaptation of the Lurgi process for use at two new gas works, at Westfield in Fife and at Coleshill, in the West Midlands. This research, and the Lurgi plants, came to be the basis of the development of substitute natural gas technology, but for town gas manufacture this line of work was overtaken by the rival possibilities, oil and natural gas.

Oil gasification was more immediately attractive, as the 1950s saw the development of the vast oil reserves of the Middle East and the construction of large refineries in Britain. The apparent prospect of endless supplies of cheap oil feedstocks encouraged the gas industry to undertake a complete transformation of its manufacturing processes. By 1968, there were only 192 works in operation, most of them very large, with the total capacity to produce 158 million m^3 of town gas a day. Of this, only 18 million m^3 of carbonizing plant capacity remained, and 2.7 million m^3 of Lurgi plant. The majority of the gas making capacity, over 115 million m^3 a day, was provided by oil-gasification processes which were almost unknown ten years earlier.

This revolution in gas manufacture produced a modern efficient industry, able to compete with other fuels on equal terms. The gas industry could offer its customers a clean, relatively economical fuel and an attractive range of appliances to go with it. Gas was thus placed in a good position to benefit from the clean air legislation which placed restrictions on the use of the traditional coal fire, and effectively ended the recurring smog which had been a major health hazard in large cities. The disappearance of the coal fire was also helped by the boom in central heating. Gas central heating was a rarity in the late 1950s, when fewer than 20,000 systems were sold each year, but by 1968 sales were near to 300,000 a year. This change in domestic habits and the rising standard of heating and comfort which people expected produced warmer homes, a cleaner environment and a revival of the gas industry. Gas stopped losing customers to electricity, and gained new ones. The industry's customers also tended to use more gas, as gas fires and central heating became the main source of heat in the home. It was an era summed up in the advertising slogan of the 1960s, "High Speed Gas".

The revolution in gas manufacture and the expansion of gas sales to the domestic market were based on the use of oil for making town gas. The third possible means of freeing the gas industry from dependence on coal carbonization, the use of natural gas, only began to play a significant part when the revival of the industry was well under way. Originally, it was thought that

108

natural gas could be used in conjunction with oil-based processes in the manufacture of town gas. A programme of exploration for natural gas on land in Britain began in 1953, but although small discoveries were made, it was thought that natural gas would have to be imported from abroad. The Gas Council therefore became involved in the world's first ocean voyage by an LNG carrier, the "Methane Pioneer", in 1959.

Having proved that importing gas as LNG was a practical possibility, the Gas Council began negotiations with Algeria, which had large gas reserves, but no domestic market for them. The first cargo of Algerian LNG arrived at Canvey Island in 1964, to be regasified and transmitted through Britain's first long-distance high-pressure pipeline. The pipeline, running 360 km from Canvey to Leeds, had connections to eight of the twelve area gas boards who would use natural gas to produce town gas.

The gas industry was barely beginning to renew itself, with oil gasification and LNG, when natural gas was found rather nearer home than Algeria. The Groningen gas field in the Netherlands was discovered in 1959, and drew attention to the possibility of finding gas in the southern basin of the North Sea, which shared the same sedimentary structure. Shell, Esso, BP and Amoco, together with the Gas Council, began geophysical surveys of the southern North Sea in 1962. The government allocated the first group of offshore production licences, allowing exploration drilling to begin, in 1964. In October 1965, BP discovered the West Sole gas field, off the Humber estuary.

This first discovery was quickly followed in 1966 by the Leman Bank, Indefatigable and Hewett fields. It was apparent that Britain had reserves of natural gas in the North Sea which would be more than sufficient to supply all the current demand for gas, if it could only be brought into use. The gas industry had customers for gas and networks of gas distribution mains, but the introduction of natural gas would mean the construction of shore terminals to receive the gas and a national system of transmission pipelines to carry the gas to the existing distribution systems. In addition, if natural gas was to be used most efficiently, it should be supplied as natural gas, with a high calorific value, rather than being reformed into town gas. The successful programme of building oil gasification plants would have to be abandoned, and every town gas burner suitable for gas with a lower calorific value would have to be converted to burn natural gas.

It was a daunting prospect, which would be far and away the biggest, most difficult and most expensive operation the gas industry had ever undertaken. Nevertheless, the Gas Council decided to take this leap in the dark, and undergo a second transformation. Less than two years after the discovery of the West Sole field, natural gas was arriving at the new shore terminal at Easington, a thousand kilometres of transmission pipeline had been constructed and conversion had begun in Burton-on-Trent.

More fields were discovered, and more gas began to flow into more terminals. The transmission pipelines were extended into all parts of Britain, so that by 1970 conversion to natural gas was in progress in all twelve areas. When conversion was at its peak, more than 2 million customers a year were undergoing the modification of all their gas appliances. It was unpopular, and there was much apprehension about the new fuel and widespread reports about the difficulty of using it. Conversion was certainly inconvenient for customers, and many problems occurred — but this was only to be expected in a programme of converting 35 million appliances in 13.5 million premises.

Conversion was completed in 1977, by which time it had cost £600 million. In addition, there was a cost of £450 million for the premature obsolescence of the works making town gas, which were now redundant. The ten years of conversion were a difficult period for the gas industry, faced with the unprecedented task and the enormous expense of building a whole new gas transmission system at the same time. In the long run, the costs and the difficulties were outweighed by the benefits to the gas industry and its customers. Gas customers gained access to a clean, non-toxic fuel, at a price much below what would now be possible for town gas; it may be an unfair comparison, but town gas in Ireland costs three times as much as natural gas in Britain. The gas industry was transformed into a modern energy-transporter, rather than manufacturer, with its future assured for many years to come by supplies of natural gas.

Over the ten years of conversion, sales of gas more than trebled. At the same time, the number of people employed in the gas industry fell from 120,000 to 100,000. The availability of natural gas on a large scale made it possible to sell gas to industry for many uses, and gas became the most popular domestic fuel. The transformation of the gas industry, from a basis of local manufacturing to a single integrated gas supply system, was reflected in the new organization set up by the Gas Act 1972. On January 1, 1973, the former Gas Council and area gas boards were merged into the British Gas Corporation, so that there was now a unified administration for the newly integrated gas industry.

110

Coincidentally, British Gas completed conversion to natural gas in September 1977, the same month as the arrival of the first gas from the Frigg field in the northern North Sea. Frigg was just the first of the new sources of natural gas, from the northern North Sea and the Irish Sea, which will allow further expansion of gas sales in the 1980s. More transmission pipelines and a new terminal will have to be built, but by the late 1970s most of the work necessary to establish an industry based on natural gas had been completed. The transition from town gas was over, and British Gas had achieved the size and structure described in this book. This natural gas industry with 14 million customers has only the most tenuous connection with William Murdoch lighting his house with coal gas, but they share the characteristics of innovation and enterprise allied to a solid engineering background, and the search for more efficient ways of making use of energy.

9

INTO THE 21st CENTURY

Forecasts

Supplies of natural gas from the North Sea amounted to some 40 billion m³ a year in the late 1970s, and were expected to rise to some 60 billion m³ a year in the late 1980s. As natural gas has a calorific value of about 39 MJ/m³, these figures are roughly equivalent to 1,600 PJ and 2,300 PJ. Making predictions about future trends in energy supply and demand is as difficult, and may be as misleading, as estimating gas reserves, but forecasts looking further ahead, perhaps into the next century, are obviously interesting, and frequently made. All forecasts depend on the assumptions of the forecasters, and assumptions of course depend on the body of knowledge presently available. In other words, forecast futures look rather like extensions of chosen characteristics of the present. Some assume that the use of energy will continue to follow its present pattern, others that the already discernible interest in energy conservation and "alternative" sources of energy will make a larger impact. However a forecast is arrived at, it cannot, by definition, predict the unpredictable — political changes, radically new technology — but must be based on what we know about the present and the immediate past.

So far as gas is concerned, demand has been rising continuously since the early 1960s, in all sectors of the energy market except transport. Its competitive price, convenience and cleanliness have helped gas to displace coal and oil in many industrial processes and in heating industrial and commercial premises. In the home, gas has become the dominant fuel: more than half the homes in Britain now have central heating, and the majority choose gas as the fuel. At the same time, British Gas is making a major effort to encourage energy conservation among its customers, to concentrate gas sales in the premium market and to develop more efficient

112

ways of using gas.

Perhaps the most authoritative forecast of the way in which we can expect the market for gas and for other fuels to develop, is that produced by the Energy Commission in 1978. It was published in a government Green Paper, "Energy Policy, a Consultative Document". It may be subject to revision from time to time, but it incorporates a number of generally accepted ideas about the use of energy in Britain up to the year 2000.

In the domestic market, energy consumption is likely to continue to rise as more and more people install central heating, and expect to live in warm, comfortable homes. The construction of new houses with much better insulation than the present norm, and increased insulation of existing houses, should lead to more efficient use of energy. The share of coal and oil will diminish. Gas will maintain its price advantage over electricity, although all prices will rise in real terms. The table shows the Green Paper forecast for the domestic market:

Domestic fuel consumption (PJ)

	1950	1970	1985	2000
Gas	147	369	918	1,055
Total	1,466	1,572	1,551	1,667

The use of energy in industry is also expected to increase, but the Green Paper suggests that the use of gas in industry may be curtailed after 1990, particularly for interruptible customers. Coal might regain some of its share of the non-premium industrial market, and gas consumption reduce to the level of firm premium markets, where the special qualities of gas are put to use, rather than gas being used in applications where lower grade fuels could do just as well. The Green Paper's forecasts are based on a faster rate of economic growth than has been the case recently:

Industrial fuel consumption (PJ)

	1950	1970	1985	2000
Gas	53	127	686	485
Total	1,983	2,585	2,796	3,724

A large proportion of Britain's energy consumption is accounted for by transport (1,182 PJ in 1970, 1,793 PJ in 2000). Oil, in its various forms, supplies nearly all the fuel used in transport, and although gas could be used, at least in theory, to power engines, there would be many practical problems. It is hardly likely that gas would be used in transport at any time, as there are many more valuable ways of using gas.

The remaining sector of the fuel market, the commercial sector, is a mixed bag of institutions such as offices, hospitals, schools, swimming pools, hotels, shops and so on. It is the smallest sector, but one of the faster growing areas of energy demand. The Green Paper forecasts that natural gas will continue to displace coal and some heating oils, but by the end of the century gas supplies to the commercial market could be limited as gas is drawn towards the domestic market:

Commercial fuel consumption (PJ)

	1950	1970	1985	2000
Gas	42	84	285	338
Total	559	781	802	950

The overall picture which emerges from these forecasts is one of a general expansion of gas sales until the late 1980s or early 1990s, as gas continues to displace coal and oil, particularly in the premium market. The effects of energy conservation, encouraged by rising energy prices, should set an upper limit to the use of gas in premium markets. The withdrawal of gas from non-premium interruptible sales, and perhaps some commercial sales, might lead to a slight decline in total gas sales by the end of the century. By then, gas will be providing a smaller proportion of total energy consumption, as gas sales are concentrated on the premium, particularly domestic, market:

Total fuel consumption (excluding transport) (PJ)

	1950	1970	1985	2000
Gas	242	580	1,889	1,878
Total	4,008	4,938	5,149	6,341

This forecast of a rise in the gas share of the energy market from 6 per cent in 1950 to 12 per cent in 1970 and 37 per cent in 1985, followed by a decline to 30 per cent in 2000, might be open to disagreement on points of detail, but it shows the kind of development which seems likely, given present knowledge of gas demand and reserves. Any number of other views is available,

114

many of them affected by the forecasters' own interests. It is hardly surprising that the electricity industry is given to predicting the imminent disappearance of North Sea gas, or that the gas industry should be more optimistic. Proponents of nuclear power predict an "energy gap" which can only be filled by nuclear power; their opponents sometimes display an unquestioning faith in alternative sources of energy, such as solar power and windmills. In other words, there are probably as many opinions as there are so-called experts.

The Green Paper is fairly conservative about the potential contribution of new technology, energy conservation and alternative energy sources. Another view was put forward in 1979 by the International Institute for Environment and Development, in its "low-energy strategy for the UK". The IIED view is generally similar to the Green Paper up to the year 2000, but puts more faith in energy conservation, so that energy consumption in all sectors except transport is rather lower, from 4,732 to 5,082 PJ, with gas having the same share of the total. Looking further ahead, the IIED predicts a significant contribution from solar power, wind generation, distributed heat and energy-saving technology by 2025, reducing total consumption to 4,339 to 5,092 PJ. Gas is still expected to supply 21 per cent of this total.

The similarities between the two forecasts are perhaps more important than the differences, which are largely a matter of opinion about the timing and impact of changes in the pattern of energy use. They agree in attributing great importance to energy conservation measures and in seeing the role of gas increasingly concentrated on premium markets. They also agree in accepting that natural gas from known discoveries will suffice to meet premium demand into the next century. If further reserves are not discovered, substitute natural gas or imported LNG will gradually supplement declining North Sea production. Forecasts looking so far into the future do not pretend to be accurate, but they help to give an idea of the way in which energy use might develop, and the continuing role of gas as far ahead as can be foreseen.

The future of gas demand

The gas industry tends to be reluctant in producing precise numerical forecasts of gas demand too far into the future, recognising that there are just too many uncertainties to take into account. Nevertheless, it is possible to pick out some of the factors likely to influence gas demand over the medium term, assuming a relatively stable political and social environment.

Firstly, gas prices will inevitably rise. Supplies of gas from the distant fields of the northern North Sea will cost more to produce than gas from the older fields in the south. The price which British Gas pays to gas producers represents only a small proportion of the cost of gas to the consumer, and its rise will be offset to some extent by the end of the need to pay for conversion and for the debt incurred in building the transmission system. There is thus likely to be a smooth gradual rise in gas prices, in line with the general rate of inflation, rather than a sudden jump. Other fuels will also be affected by inflation, so the price of gas relative to its competitors will remain much the same. Rising prices will not act as an incentive to switch to another fuel, but they should help to convince gas users of the desirability of saving energy.

There is still much to be done to reduce energy consumption in existing houses — about a third of all hot water storage tanks have no lagging to keep the heat in, and half of all houses have little or no roof insulation. These easy ways of saving energy pay for themselves very quickly, and with a little help from government grants or loans, should become more common. Much more extensive savings are possible in houses designed with low energy use in mind. As yet, low energy houses are rare except as experiments, but it could soon become normal practice to design new houses with heavy insulation, double glazing and possibly some form of solar water heating for use in suitable weather. The impact of such new houses can only be gradual, as existing houses will obviously be lived in for many years, but it does seem likely that energy conservation will become a priority with architects when they are designing new houses.

The gas appliances used in well-insulated buildings will have to be designed to operate at high efficiences at lower heat outputs than are now the norm. This should continue the trend towards more efficient appliances which is already apparent. Heat consumption in British homes is much the same as it was twenty years ago, despite a great improvement in the comfort standards which people expect and the building of more houses. The same amount of heat is being put to better use, as it is supplied in a more efficient form — mainly, gas rather than coal — and used in better designed appliances. Cookers, for instance, are no longer sold with pilot lights which burn all the time, but with spark ignition. Radiant convector gas fires, with heat exchangers to produce warm air, heat a room more effectively than the old radiant-only fires. The control of central heating has become more precise, so that it delivers just as much heat as is required, when it is needed. Such

116

improvements should continue, with new appliances progressively replacing less efficient old ones.

The fact that buildings are going to be designed to use less energy, and that even in old buildings insulation and the natural renewal of appliances will contribute to energy saving will not lead to an immediate reduction in gas demand. Insulation does not cause people to use less fuel to stay at the same level of comfort; instead, they choose to live in warmer homes, so that the energy savings from insulation are frequently only half of what might have been expected. There is plenty of scope for expansion of gas sales, both among existing users and with new customers. Central heating, the domestic use which consumes large quantities of gas, first reached 50 per cent penetration of the total number of homes as recently as 1977, so sales can continue for some years to first-time buyers. Of the 19 million homes in Britain, 14 million have a gas supply, and a further 2½ million could easily be connected to a gas main, if they want gas.

The gain of new gas customers and the spread of central heating should result in a continuing growth of gas sales to the domestic market, from about 750 PJ in 1979 to 1,050 PJ in the late 1980s. By then, the gas share of domestic heat supply should have grown from 45 to 60 per cent. This represents saturation level, beyond which growth will just not be possible, as some areas will remain forever without gas and some domestic energy uses like lighting will naturally be supplied by electricity. By this stage, energy saving measures will be starting to have a perceptible effect, so a plateau in domestic gas demand will be reached, perhaps to be followed by a slow decline.

A similar pattern is likely to become apparent in the commercial market and in the premium industrial market, affected by the conflicting forces of growing demand for gas as the most economical fuel and more stringent energy saving measures propagated by British Gas. As these markets are also subject to intense competition from other fuels, and many users can change fairly easily from one fuel to another as the price advantage changes, the outcome is probably beyond prediction. What is certain is that British Gas does not intend to expand sales to non-premium users, preferring to reserve gas as far as possible for those applications to which gas is best suited. Generally, British Gas has moved away from the situation which prevailed when supplies from the southern North Sea were coming ashore in unprecedented quantities and had to be disposed of, even to non-premium users. In future, British Gas intends to use growing supplies from the North Sea to meet growth in premium demand. Supply will follow demand, rather than supply dictating how much gas is to be sold.

The future of gas supplies

This policy of allowing demand to dictate the amount of gas produced, rather than just producing as much gas as possible as quickly as possible, has already begun to take effect. Many of the contracts between British Gas and the producer companies have been renegotiated, to allow more flexibility in the amount of gas British Gas purchases and to encourage the production of all the gas that is recoverable in the long term. As more gas becomes available from the fields in the northern North Sea, production in the south can be restricted to the level needed to meet demand.

It has already been mentioned that supplies from the southern North Sea reached a level of about 40 billion m^3 in 1977. With the addition of some 15 billion m^3 a year from Frigg at its peak rate, over 6 billion m^3 a year from Brent and smaller supplies from several of the northern oil fields with associated gas, 60 billion m^3 a year should easily be available in the late 1980s even if some southern fields are then in decline. By then the Morecambe field will also be contributing to our gas supplies — if all goes well, the terminal on the Irish Sea coast could be commissioned in 1984.

In other words, supplies of gas from fields which are known and under contract to British Gas will be able to meet the likely rise in demand to its plateau rate in the late 1980s. Beyond then, supplies from these sources will be entering a long slow decline; at the same time, demand could well be diminishing. As British Gas intends to concentrate gas sales on the premium markets, particularly the domestic sector, there should be a loss of demand from non-premium users, while many premium users will be benefiting from energy conservation measures. The net result is that premium customers are assured of a gas supply into the next century from the fields which are already known.

This forecast of course takes no account of sources of gas beyond those which are already known and assured. It is possible that much more natural gas could become available, as exploration opens up areas which have so far been ignored. The exploration of new areas, and secondary exploration in areas which already produce gas, is very likely to find more gas. The past record of success in finding so much gas in a relatively short period gives reasonable grounds for optimism. In addition, it should not be forgotten that gas need not only come from the British sector of
118

the North Sea, and that there are large known gas reserves in the Norwegian sector, in fields like Statfjord. Statfjord has been discovered, but no decision has been made on how to develop it. It is open to British Gas to buy gas from Statfjord, and from any other fields which might be found in the Norwegian sector, which could be delivered to Britain like Frigg gas.

These potential additional offshore supplies cannot of course be guaranteed. Even if they do become available, there will still come a time when there will be a need to supplement gas from the North Sea. They would help to defer that time, which with presently contracted gas resources is likely to arrive in the late 1990s, but sooner or later it will be necessary to find means of providing more gas to meet peak demand in winter. Gradually, the North Sea will be able to supply less and less of gas demand, and supplementary supplies will become increasingly important, eventually replacing North Sea gas.

Two obvious supplementary sources of gas have already been mentioned, imported LNG and the manufacture of substitute natural gas (SNG). It is difficult to say how much LNG will be available by the end of the century, as so much depends on the rate of development of natural gas reserves in remote countries. It does seem probable that, with oil and natural gas production in decline in many of the traditional producing areas and LNG being more widely traded, there will be intense competition between the industrialized countries to purchase whatever LNG is being produced.

Britain could retain a greater degree of self-sufficiency by using SNG, made from coal. British Gas has already developed and demonstrated the technology for making SNG at its pilot plant at Westfield, in Scotland. SNG can be made from coal with a thermal efficiency approaching 70 per cent, compared with little over 30 per cent efficiency in coal-fired power stations. The provision of heat from coal via SNG is thus clearly competitive with the use of electricity as a means of conveying energy to the consumer. SNG would be interchangeable with natural gas from the North Sea, so there would be no need to convert appliances, and the existing system of transmission pipelines and distribution mains could continue to be used.

There is said to be enough coal in Britain to meet demand for hundreds of years, so there would be no problem about the raw material for SNG manufacture. About thirty very large SNG plants would be enough to meet all Britain's needs. It would be unwise to attempt to predict what competition gas might be facing from

119

other sources of energy in the 21st century: the promise of cheap nuclear power might be fulfilled with fusion reactors, solar power and the generation of electricity by windmills and tidal power could make some kind of contribution. Low energy housing and such developments as the heat pump could produce considerable reductions in demand if they became the norm. Nevertheless, there would still presumably be a demand for gas in its traditional role as a fuel which provides clean controllable heat when and where it is needed.

An industry mainly concerned with making and supplying SNG could continue almost indefinitely to provide enough gas to meet premium demand. How big that demand will be in, say, a hundred years from now is beyond anyone's power to predict. The only thing we know for sure about the future is that it will be unimaginably different from the present, but whatever sort of society exists in the next century, it will need energy. Gas will be there to meet this need.

APPENDIX
NUMBERS AND UNITS

Billions and trillions

The American meanings of billion and trillion have been adopted by the petroleum industry throughout the world, and are now used in Britain in government publications and in most newspapers. This book follows what now seems to be normal practice: billion means thousand million (10^9); trillion means million million (10^{12}).

Metric and imperial units

The gas industry in Britain has traditionally used imperial units to measure quantity, distance and pressure, and the law which governs most of its activities, the Gas Act 1972, prescribes the continued use of the cubic foot, the therm and the British thermal unit. British Gas must charge for gas on the basis of its heat content, according to the number of therms supplied. The number of therms must be calculated on the basis of the calorific value of the gas, which is defined as the number of British thermal units produced by the combustion of one cubic foot of gas, measured at 60°F under a pressure of 30 inches of mercury. In its turn, the British thermal unit is defined as the quantity of heat necessary to raise the temperature of one pound of water at 15°C by 1°F (sic). Finally, one therm equals 100,000 British thermal units.

There is a clause in the Gas Act which enables the government to decree that the gas industry should adopt metric units. As yet, the government has not seen fit to do so, but when it does the volume of gas will presumably be measured in cubic metres and its heat content in multiples of the joule. Many readers of this book will be more familiar with SI than with obscure imperial units like the British thermal unit. North Sea gas production is now often measured in cubic metres, and SI will surely take over soon, so this book uses metric units throughout.

Conversion factors

Volume:
1 cubic metre (m^3)	= 35.993 ft^3
1 cubic foot (ft^3)	= 0.02778 m^3

The cubic metre is measured under metric standard conditions, dry. The cubic foot is measured under imperial standard conditions, saturated.

Heat:
1 joule (J)	= 0.000948 Btu
1 British thermal unit (Btu)	= 1055.06 J
1 megajoule (MJ)	= 947.817 Btu
1 therm = 100,000 Btu	= 105.506 MJ

The above conversion factors for the Btu and hence the therm are based on the International Table Btu, which British Gas uses in contracts. A slightly different value applies to the Btu as defined by the Controller of Gas Standards (1 Btu = 1054.73 J) for calorific values.

Incidentally, the joule is a very small unit, so large quantities of heat are measured in gigajoules (GJ, 10^9 joules), terajoules (TJ, 10^{12} joules) and petajoules (PJ, 10^{15} joules).

Calorific value:
1 MJ/m^3	= 26.34 Btu/ft^3
1 Btu/ft^3	= 0.03796 MJ/m^3

Pressure:
1 bar	= 100,000 Pa	
1 bar	= 0.987 atm	= 14.5038 lbf/in^2
1 std. atmosphere (atm)	= 1.01325 bar	
1 millibar (mbar)	= 0.40146 inH_2O	= 0.0145 lbf/in^2
1 inch of water (inH_2O)	= 2.4909 mbar	
1 pound-force per square inch (lbf/in^2 or p.s.i.)	= 68.9476 mbar	

The SI unit for pressure, the pascal (Pa), is seldom used. The International Gas Union recommends the use of the supplementary metric units, the bar and millibar, which is the practice generally followed by British industry.

Calorific values

Both town gas and natural gas are mixtures of various gases, in various proportions. Their compositions, and thus their calorific values, differ according to the process used to make town gas and to the field from which natural gas is produced. The principal combustible constituents of town gas and natural gas have the following approximate calorific values:

	Btu/ft^3	MJ/m^3
carbon monoxide (CO)	315	12.0
hydrogen (H$_2$)	318	12.1
methane (CH$_4$)	992	37.7
ethylene (C$_2$H$_4$)	1,572	60.0
ethane (C$_2$H$_6$)	1,738	66.0
propane (C$_3$H$_8$)	2,473	93.9
butane (C$_4$H$_{10}$)	3,200	121.5
benzene (C$_6$H$_6$)	3,678	139.6

The typical town gas which used to be manufactured in Britain had a calorific value of 500 Btu/ft^3 or 19.0 MJ/m^3. A fairly typical natural gas mixture, as supplied to customers now, would have a calorific value of 1,023 Btu/ft^3 or 38.9 MJ/m^3.

FURTHER READING

Books specifically about gas and the gas industry suitable for the student and the general reader soon become out of date. Two of the best such works, both written by D. Scott Wilson, "The Modern Gas Industry" (1969) and "North Sea Heritage: The Story of Britain's Natural Gas" (1974), present a useful guide to the gas industry at those dates, but are now out of print.

The only work about gas for the general reader currently available is a Science Museum booklet, "Gas: An Energy Industry" (HMSO, 1976) by Susan E. Messham.

For the science student, there is "Gasmaking and Natural Gas" (1972), published by BP, and "Natural Gas: A Study" (1972) by E.N. Tiratsoo.

Perhaps the best way of keeping up to date with what goes on in the gas industry, particularly its sales and finances, is the British Gas Corporation "Annual Report and Accounts", published each year by HMSO. It includes a useful statistical section.

The main statistical source for all forms of energy is the "Digest of UK Energy Statistics", produced by the Department of Energy and published annually by HMSO. The main statistical series are kept up to date in the Department of Energy's "Energy Trends: A Statistical Bulletin", available monthly from the Central Office of Information.

The chief source of factual information about offshore oil and gas is the Department of Energy's "Development of the Oil and Gas Resources of the UK", generally known as the Brown Book, published annually by HMSO.

"Energy Policy: A Consultative Document" (HMSO, 1978), better known as the Green Paper, discusses energy supply and demand in Britain up to the year 2000.